The Far Western Frontier

The Far Western Frontier

Advisory Editor

RAY A. BILLINGTON

Senior Research Associate
at the Henry E. Huntington Library
and Art Gallery

MINERS AND TRAVELERS'

GUIDE

TO

OREGON, WASHINGTON, IDAHO, MONTANA, WYOMING, AND COLORADO.

VIA

THE MISSOURI AND COLUMBIA RIVERS

PREPARED BY

CAPTAIN JOHN MULLAN

ARNO PRESS

A NEW YORK TIMES COMPANY

New York • 1973

Reprint Edition 1973 by Arno Press Inc.

Reprinted from a copy in The State
Historical Society of Wisconsin Library

The Far Western Frontier
ISBN for complete set: 0-405-04955-2
See last pages of this volume for titles.

Manufactured in the United States of America

Publisher's Note: The "General Map of the North
Pacific States and Territories" has been reproduced
from the best available copy.

Library of Congress Cataloging in Publication Data

Mullan, John, 1830-1909.
    Miners and travelers' guide to Oregon.

    (The Far Western frontier)
    Reprint of the 1865 ed.
    1. Northwest, Pacific. I. Title. II. Series.
F852.M95 1973   917.95'04'4   72-9461
ISBN 0-405-04989-7

MINERS AND TRAVELERS'

GUIDE

TO

OREGON, WASHINGTON, IDAHO, MONTANA, WYOMING, AND COLORADO.

VIA

THE MISSOURI AND COLUMBIA RIVERS.

ACCOMPANIED BY

A GENERAL MAP OF THE MINERAL REGION OF THE NORTHERN SECTIONS OF THE ROCKY MOUNTAINS.

PREPARED BY

CAPTAIN JOHN MULLAN,

LATE SUPERINTENDENT OF THE NORTHERN OVERLAND WAGON ROAD, AND COMMISSIONER OF NORTHERN PACIFIC RAILROAD.

NEW YORK:
PUBLISHED BY WM. M. FRANKLIN,
(FOR THE AUTHOR),
24 VESEY STREET,
1865.

C. A. ALVORD, ELECTROTYPER AND PRINTER.

CONTENTS.

MINERS AND TRAVELERS'
GUIDE

THE following pages and accompanying map have been prepared with a view to place in the hands of travelers and emigrants to the North-West and Pacific Coast, such advice and information as they would find useful in their journey, when traveling by the new route.

During the past fourteen years the entire emigration that has sought the Pacific Coast, with the view of there making permanent homes, has taken either the route across the Con tinent, via the South Pass, involving 2000 miles of land travel, or via the Isthmus of Panama, involving 6000 miles of sea travel ; both fraught with heavy expense, danger and discomfort.

The great desire of all has been to secure a route where the sea travel would be avoided in toto ; and at the same time have the land transit the shortest minimum. The geography of the section of the Continent west of the Mississippi shows that this can only be attained by ascending the Missouri river to its highest point practicable for steamers, and thence cross to the navigable waters of the Columbia, where we find the land carriage only 624 miles.

Having been occupied for a number of years in the exploration and construction of the wagon road, via this route, I feel warranted in placing in a brief form such advice, facts, and statements as our labors in the field have developed.

Those who desire to make this trip should apply for further information to Charles P. Chouteau, of St. Louis, or to John G. Copelan, of St. Louis, both of whom are interested in forwarding passengers and freight from St. Louis to the Rocky Mountain region, at the headwaters of the Missouri and Columbia Rivers. Their steamers are generally ready to leave St. Louis somewhere between the 4th of March and 1st of May—starting thus early in order to take advantage of the June rise which they meet at or near Sioux City, and which enables them to run over all the bars and shoals found in the difficult stream of the Missouri.

These boats make the trip but once a year, and hence all travelers should make their preparations in time to take the boats by 1st April, either from St. Louis or from Walla-Walla. John G. Copelan will keep constantly a steamer between Fort Benton and the Yellowstone for the accommodation of travelers who wish to return east in early spring, or late in the autumn. Wagons and outfits of all kinds can at present be secured at Fort Benton.

Travelers will probably find fresh vegetables at Sun River, on the Big Prickly Pear, in the Deer Lodge Valley, Hellsgate Valley, at the Cœur d'Aléne Mission, on the Spokane River, on the Touchet River, Dry Creek and Walla-Walla, and fresh beef at each and all of these points.

Fresh animals can be purchased at nearly all these points,

and blacksmith shops will be found at the Deer Lodge,
Hellsgate and Cœur d'Aléne Mission.

No fear need be apprehended from Indians along the en
tire route. The trip from St. Louis to Fort Benton will
involve from 35 to 40 days, and from Fort Benton to Walla-
Walla about the same length of time.

The following more detailed statistics published in an of-
ficial report, will be found to contain much of interest to
those who have never made the trip :

RECOMMENDATIONS FOR TRAVELERS

For persons who desire to leave St. Louis in the spring
on steamer for Fort Benton, where the passage is from $100
to $200, and freight from ten cents to twelve cents per pound,
and who desire to make the land transit by wagon, I would
advise that they provide themselves with a light spring covered wagon in St. Louis, also two or four sets of strong har-
ness, and transport them to Fort Benton, where they can
procure their animals, mules or horses. The former can be
had from $100 to $150, the latter from $50 to $75; oxen,
from $100 to $125 per yoke. Let them provide themselves
with a small kit of good strong tin or plated iron mess fur-
niture ; kettles to fit one in the other, tin plates and cups,
and strong knives and forks ; purchase their own supplies
in St. Louis ; brown sugar, coffee, or tea, bacon, flour, salt,
beans, sardines, and a few jars of pickles and preserved
fruits will constitute a perfect outfit in this department. I
have found that for ten men for fifty days, the following is
none too much on a trip of this kind : 625 pounds of flour,
50 pounds of coffee, 75 pounds of sugar, 2 bushels of beans,

1 bushel of salt, 625 pounds of bacon sides, 2 gallons of vinegar, 20 pounds of dried apples, 3 dozen of yeast powders, and by all means take two strong covered ovens, (Dutch ovens.) These amounts can be increased or diminished in proportion to the number of men and number of days. If your wagon tires become loose on the road, caulk them with old gunny sacks, or in lieu thereof, with any other sacking; also, soak the wheels well in water whenever an opportunity occurs. In loading the wagons, an allowance of four hundred pounds to the animal will be found sufficient for a long journey. For riding saddles, select a California or Mexican tree with machiers and taphederos, hair girth, double grey saddle blanket, and strong snaffle bit.

If the intention is to travel with a pack train, take the cross-tree packsaddle, with crupper and breeching, and broad thick pads. Use lash-rope, with canvas or leather belly bands. Have a double blanket under each saddle. Balance the load equally on the two sides of the animal—the whole not to exceed two hundred pounds. Have a canvas cover for each pack. A mule-blind may be found useful in packing. Each pack animal should have a hackama, and every animal (packing and riding) a picket-rope, from thirty-five to forty feet long and one inch in diameter. For my own purposes, I have always preferred the apparejo for packing, and have always preferred mules to horses. Packages of any shape can be loaded upon the apparejo more conveniently than upon the packsaddle. A bell animal should be always kept with a pack train, and a grey mare is generally preferred. Every article to be used in crossing the plains should be of the best manufacture and strongest material. This will, in the end, prove true economy. Animals should be shod on the fore-feet, at least. Starting at dawn and camping not later than 2 P. M., I have always found the best plan in marching. Animals should not go out of a walk or a slow trot, and after being unloaded in camp they should always be allowed to stand with their saddles on and girths loose, for at least fifteen minutes, as the sudden exposure of their warm backs to the air tends to scald them. They should be regularly watered, morning, noon, and night. Never maltreat them, but govern them as you would a woman, with kindness, affection, and caresses, and you will be repaid by their docility and easy management. If you travel with a wagon, provide yourself with a jackscrew, extra tongue, and coupling pole ; also, axle-grease, a hatchet and nails, auger, rope, twine, and one or two chains for wheel locking, and one or two extra whippletrees, as well as such other articles as in your own judgment may be deemed necessary. A light canvas tent, with poles that fold in the middle by a hinge, I have always found most convenient. Tables and chairs can be dispensed with, but if deemed absolutely necessary, the old army camp stool, and a table with lid that removes and legs that fold under, I have found to best subserve all camp requisites. Never take anything not absolutely necessary. This is a rule of all experienced voyageurs.

ADVICE TO EMIGRANTS BY THIS ROUTE.

Those who start from the Upper Mississippi frontier can replenish their supplies at Fort Union, at the mouth of the Yellowstone, at Fort Benton, and, in addition, at the other points hitherto alluded to.

Those who travel by the central or Platte route, and desire to take the western section of the road to Walla-Walla, can deflect either at Fort Laramie, the Red Buttes, or Fort Hall, and connect with it at the Deer Lodge Valley.

The road from Fort Laramie to the Deer Lodge Valley has never been worked, but was passed over with wagons by Captain Reynolds, of the army, in 1859 and 1860, and by miners in 1863–64. It passes through a beautiful, easy, and interesting region.

The road from Fort Hall to Deer Lodge has been used by wagons for many years, and though not worked is quite practicable.

ITINERARY OF ROUTE.

The following Itinerary supposes the traveler to start from Walla-Walla; but by simply reversing the order of the record, no trouble will be had:

First day.—Leave Walla-Walla and move seven and a half miles, to Dry Creek, and encamp at crossing; easy rolling prairie hills en route; wood, water, and grass at camp.

Second day.—Leave Dry Creek and move eleven and a half miles, to Touchet Bridge, and encamp; easy rolling prairie hills en route; wood, water, and grass abundant.

Third day.—From Touchet take wood in wagons for two days; move seven miles, to the springs, and encamp; grass and water here, but no wood; level prairie road en route.

Fourth day.—Leave springs and move to Snake River; distance, twenty miles; grass, water, and drift wood here; graze animals on hills on left bank; good road over rolling prairie, somewhat hilly.

Fifth day.—Cross Snake River by ferry-boat; charge for wagons, $4; men, fifty cents each; riding and pack animals, fifty cents each; swim loose stock, or, if preferred ferry same. Move to Palouse; distance, fourteen and a half miles; water and grass; willows for fuel. It would be well to take a small quantity of drift wood along from Snake River; good road.

Sixth day.—Move to Cow Creek; distance, eleven miles. Wood, water, and grass at camp; good place to rest animals for a day, if required.

Seventh day.—Move to Aspen Grove; distance, 18 miles; good road. Wood, water, and grass at camp; good place to rest animals a day, if required.

Eighth day.—Move to Lagoon camp; distance, twentyone and a half miles; good road; wood, water, and grass at camp.

Ninth day.—Move to Rock Creek; distance, twelve and a half miles; somewhat stony, but animals should be shod, in which case they will travel well; wood, water, and grass at camp.

Tenth day.—Move to Hangman's Creek; distance nineteen miles; good road; wood, water, and grass at camp.

Eleventh day.—Move to Spokane River; distance, twelve and a half miles, and cross; wood, water, and grass at camp; good place to rest animals; charge for each wagon, $4; for each man, fifty cents; swim loose stock, or ferry, if preferred. There is a ford eight miles above.

Twelfth day.—Move to camp on Spokane, at the edge of the timber; distance, sixteen miles; good road; wood, water, and grass abundant.

Thirteenth day.—Move to Wolf's Lodge Prairie; distance, eighteen miles; road hilly in places, but not bad; wood, water, and grass at camp.

Fourteenth day.—Move to the Cœur d'Aléne Mission; distance, seventeen miles; road hilly at one or two places, but not bad; wood, water, and grass at the mission. Good place to rest animals for a day or two, and which is by all means advisable, as you now enter the timber, where camp-grounds have to be specially selected, and the animals should be well rested. Vegetables may be had at the mission.

Fifteenth day.—Move to Mud or Ten-mile Prairie; road good; in very early spring may be wet in places; good camp for wood, grass, and water. A good camp may also be had in seven miles from the mission, in open timber; water three hundred yards to the north of road, in running creek; good place for animals.

Sixteenth day.—Move to the fifteenth bridge; good water and wood; grass is not very abundant, but there are a number of small prairies above and below this bridge where grass is found; about half a mile below the bridge, on right bank, is a fine prairie; road good; distance, sixteen miles.

Seventeenth day.—Move to Johnson's Cut-off, which is a ravine from the north. The head of this ravine and the hills around it furnish an abundance of grass. This *may be* the worst day's march, as it involves many crossings, and the road may be wet; distance, eleven miles. An endeavor should be made to camp here at the risk of getting late into camp.

Eighteenth day.—Move to Long Prairie; distance, three miles; road good, unless during the freshet, when some of the crossings may be swollen. Long Prairie is one mile long, one fourth of a mile broad; grass in large portion; grass also on hills to its north, just before the descent into the prairie; a blind trail leads to it through the timber.

Nineteenth day.—Make an early start and cross summit of Bitter Root mountains; may have to double teams at second curve. Move to the Five-Mile Prairie, on St. Regis Borgia River; distance, eleven and a half miles; grass sparse; wood and water abundant.

Twentieth day.—Move eleven and a half miles to Sawyer's Prairie; wood, water, and grass abundant; road good.

Twenty-first day.—Move to Cantonment Jordan, five and a half miles distant; grass half a mile above camp; wood and water everywhere.

Twenty-second day.—Move to Bitter Root Ferry; distance, fourteen and a quarter miles; wood, water, and grass abundant.

Twenty-third day.—Cross ferry and go to Seven-Mile Prairie; charges for crossing the same as at the Spokane and Snake Ferries. The stream is fordable in very low water, but I would advise all strangers to cross in the ferry-boat, as the ford is a dangerous one, except to those who know it well. Rest your animals at this point. Good camp, with wood, water, and grass.

Twenty-fourth day.—Move nine and a half miles to Brown's prairie; good road; wood, water, and grass at camp.

Twenty-fifth day.—Move fifteen and a half miles to west end of Big Side-cut, and camp at foot of mountain, on small creek. Wood, water, and grass abundant; may have to double teams over Brown's Cut-off divide, going either way; road good, with this exception.

Twenty-sixth day.—Move over Big Side-cut to a camp on

Main River, one mile above the Rocky Points, where the road passes through a rocky defile; distance, seventeen miles; road fair; wood, water, and grass at camp.

Twenty-seventh day.—Move to Brown's house, twelve miles distant; road good; wood, water, and grass at camp.

Twenty-eighth day.—Move to Higgin's and Worden's store, at Hell's Gate, distance twelve and a half miles; road excellent; wood, water, and grass here; good place to rest animals for a day or two; blacksmith's shop at Van Dorn's, and supplies of all kinds can be obtained, dry goods, groceries, beef, vegetables, and fresh animals, if needed.

Twenty-ninth day.—Move to Big Blackfoot bridge, eleven miles; road good; wood, water, and grass abundant.

Thirtieth day.—Move to Campbell's camp, fifteen miles; excellent road; good wood, water, and grass abound.

Thirty-first day.—Move to Lannon's camp, nine miles; road excellent; may have to double teams at Beaver Tail butte; wood, water, and grass abundant.

Thirty-second day.—Move eleven miles to Lyon's Creek, crossing en route Hell's Gate bridge; road good; wood, water, and grass at camp.

Thirty-third day.—Move to Flint Creek, distance eleven miles; road somewhat hilly but still not steep; wood, water, and grass at camp.

Thirth-fourth day.—Move thirteen and one half miles to Gold Creek or American Fork of Hell's Gate River; road excellent; wood, water, and grass at camp; supplies of all kinds to be had here, dry goods, groceries, fresh beef, animals, and possibly vegetables.

Thirty-fifth day.—Move to Deer Lodge River, distance sixteen miles; road hilly but not requiring double teaming; wood, water, and grass at camp; Deer Lodge would be found a good place to rest for a day or two; fresh beef to be had here from the settlers.

Thirty-sixth day.—Move to Little Blackfoot River, seventeen and a quarter miles; road generally good, hilly at one or two points, but not steep; good wood, grass, and water at camp.

Thirty-seventh day.—Move to west base of Rocky Mountains, at Mullan's Pass, thirteen and a half miles; road generally good, though sometimes wet in early spring; no ascending the north fork of Little Blackfoot; wood, water, and grass at camp.

Thirty-eighth day.—Cross summit of Rocky Mountains and go seven miles to Fir Creek; road good; wood, water, and grass at camp.

Thirty-ninth day.—Move seventeen miles to Little Prickly Pear Creek; road hilly in places but not bad; camps at shorter distances can be made, as several creeks are passed en route; Soft Bed Creek midway offers a good camp; this would be a good place to rest animals.

Fortieth day.—Start early and go over Medicine Rock Mountain fifteen and a half miles; this is the worst day's march; road rocky in places, but, with care, easily made; wood, water, and grass at camp on Oversight Creek.

Forty-first day.—Move twenty miles to the Dearborn River; wood, water, and grass at camp.

Forty-second day.—Move to Bird Tail Rock, fifteen miles; road excellent; water and grass at camp; willows for fuel but scant; it would be well to pack wood from the Dearborn or Sun Rivers, according to which way you are traveling.

Forty-third day.—Move to Blackfoot Agency, or Sun River, eighteen and a half miles; excellent road; wood, water, and grass at camp; good place to rest animals for a day or two; in high water there is a ferry-boat for crossing.

Forty-fourth day.—Move eight miles to the point where you leave Sun River; road excellent; wood, water, and grass at camp.

Forty-fifth day.—Move sixteen miles to the lake; road excellent; water and grass; take wood from camp.

Forty-sixth day.—Move to the springs, seven miles; water and grass, but no wood; or you can go to the Big Coulée, sixteen miles further, and encamp on the Missouri River; road good.

Forty-seventh day.—Move to Fort Benton, twenty-seven miles, if you encamp at the springs, or eleven miles if you encamp at the Big Coulée. The latter never was a portion of my road, but was worked by Major Delancy Floyd Jones, and I am not responsible either for its location or the character of the work performed.

If you are going from Fort Benton, it would be preferable to camp at the spring. This can be accomplished by starting early; and I should advise all parties traveling with wagons to avoid the Big Coulée. If water be not sufficiently abundant at the springs, then encamp at the lake. It may be found best to start at dawn and make the lake. The road is excellent.

The total distance herein given is six hundred and twenty-four miles, made in forty-seven days traveling; or, allowing eighteen days for delays, contingencies, and recruiting animals, in fifty-five days, with loaded wagons; or in thirty five days if you are traveling with pack animals.

GENERAL DIVISION OF ROUTE.

Our road involved one hundred and twenty miles of difficult timber-cutting, twenty-five feet broad, and thirty measured miles of excavation, fifteen to twenty feet wide. The remainder was either through an open, timbered country, or over open, rolling prairie. From Walla-Walla eastward the country might be described in succinct terms as follows: First one hundred and eighty miles, open, level, or rolling prairie; next one hundred and twenty miles, densely timbered mountain bottoms; next two hundred and twenty-four miles, open timbered plateaus, with long stretches of prairie; and next one hundred miles, level or rolling prairie. Thus it is seen that the Rocky and Bitter Root Mountains rise midway in our route, with long prairie slopes on either side; that the latter are intersected in every direction by streams flowing from both water-sheds, and rising in the heart of the mountain system; that these prairie stretches interpose but slight obstructions to the location of a road, and it is only in the more elevated central sections where our sterner engineering problems are to be met.

INDIANS ALONG THE ROAD.

The Indians met with along the line are the the Palouse, Spokane, Cœur-d'Aléne, Flatheads, Pend-d'Oreilles, a few Snakes and Bannocks, the Blackfeet, and the mountain Nez Percés.

The Walla-Wallas and the Cayuse have all been removed to the agency on the Umatilla.

The Palouse number about two hundred, reside on the

Banks of the Snake and Palouse Rivers, and live solely by
fishing. I do not know that they cultivate the soil. The
absence of any great amount of farming land in their country has always possibly prevented them from attempting it.

They have no treaty arrangements with the government,
and I think they could be assigned to the Nez Percés reservation with advantage to the government and security to
themselves. They are miserable creatures; have neither
houses nor lodges, but live under wicker shelters. They
own very few horses. The Spokanes number about five
hundred; live by fishing and cultivating small patches of
land. They reside, at times, on the Spokane, at times on
the Spokane plains, have lodges and houses, and are superior
to the Palouse. They have no treaty with the government,
and they might, with the Colville Indians, be located near
the military post of Fort Colville. They are friendly when
it is to their interest.

The Cœur-d'Alénes number about three hundred, live at
the mission, and along the Cœur-d'Aléne and St. Joseph's
rivers. They own houses, cattle, and canoes, and with the
Spokanes and Nez Percés often cross the mountains in quest
of buffalo. They live by hunting, fishing and cultivating
the soil. They have no treaty with the government, and I
think they should be moved to the Flathead reservation;
they live partly in log-houses, mostly in skin lodges.

The Flatheads number about four hundred, and live by
hunting and cultivating the soil. They are the best Indians
in the mountains. They have treaty arrangements with the
government, but have never gone upon their reservation on
the Jocko River; no steps have ever been taken to remove
them thence, and they still reside in St. Mary's Valley, which,
by the terms of the treaty, was guaranteed to them. They

think the government has not kept its faith in not confirming this valley to them. Under judicious management they and the Pend-d'Oreilles might be made to go upon a joint reservation, contemplated in a treaty made with them by Governor Stevens. They own great numbers of horses and cattle, and cultivate the soil more than any Indians except the Pend-d'Oreilles. They are friendly, and under their chiefs, Victor, Ambrose, and Moïse, will always remain so unless some great injustice is done them. They and the Pend-d'Oreilles go annually to the buffalo hunt on the plains of the Missouri. They live partly in houses, mostly in skin lodges.

The Pend-d'Oreilles reside principally at the Pend-d'Oreille mission, live by hunting, fishing, and cultivating the soil ; as a tribe they are friendly, though there are some bad fellows among them. They number about five hundred souls, and have treaty arrangements with the government, though they have never moved on to their reservation.

The mountain Nez Percés number from one to two hundred, live and hunt with the Flatheads, and are an annoyance both to them and the whites. They should be either incorporated with the Flatheads or made to live with their tribe. They are generally disaffected, and cause much trouble and disturbance in the country.

With the Flatheads are found a few Indians of the Iroquois, Shawnee, and Snakes, and one or two New Mexican Indians. They all find a friendly and welcome home with the Flatheads, into which tribe they have married.

Occasionally a few Snakes and Bannocks come to the Deer Lodge Valley. They live generally in the Beaver Head and on the Salmon River. I do not know their number or condition. I only know they are adept horse thieves, have no treaty arrangements with the government, and need to be looked after both for the security of the frontier settlements and their own good.

I especially invite the attention of the Indian Department to the necessity of having some arrangements with the Snakes, Bannocks, and Spokanes, and point the Beaver Head out as a suitable point for collecting them on a general reservation, where a large military post should be established to keep them in order.

The Blackfeet number from eight to ten thousand souls, and live exclusively by hunting the buffalo. They live partly in our territory, to the north of Fort Benton, and partly in British territory. They have treaty arrangements with the government, and, in the absence of military force to control them, keep their faith as well as could be expected from wild savages. They are rich in horses and wives, for they are perfect Mormons in polygamy—all the other tribes practice monogamy. They are great horse thieves, though I never suffered from this propensity, to which they are greatly addicted. As a people, these Indians have as high a regard for the rights of *meum and tuum* as their superiors, the whites; and if their true condition was known at the Indian bureau, I am sanguine an improvement for the better would take place. With the present system at the bureau, however, I can only expect to see experiments and changes made uutil the Indian has disappeared.

The present Superintendent of Indian Affairs in Washington Territory, Mr. C. E. Hale, has instituted a project that I have long indorsed, and the only one likely to save any portion of the Indian tribes. This is to take the children and educate them under a proper system; for it is as difficult to mould the ideas and acts of an Indian, after he has passed the age of twenty-one, as it is those of his white neighbor ; and it is only by taking the children, and rescuing them from sloth, ignorance, and savage propensities, that any decided improvement can be attained.

For myself I should like to see the supervision of the Indians transferred to the War Department, so that the hand that rewards should be the one to punish when needed, and thus produce a more uniform and harmonious management, giving greater security against outbreak, and a more economical administration of the finances of the government. If this cannot be done, of which I despair, since it aims a blow at executive patronage, then I should by all means advocate that the Superintendent of Indian Affairs be allowed to appoint his own agents. This is as natural as it is just ; the agents should be directly responsible to the superintendent whom the government has charged with their general supervision, and this supervision cannot be properly maintained when the agents receive their authority from another and different source, and are thus inclined to slight, if not ignore, his authority.

The sphere of duties of the superintendents should never be so great as to prevent them from visiting every agency once a year ; in Oregon and Washington this is impracticable ; and I would, therefore, recommend the establishment of a Rocky Mountain superintendency, with its headquarters in the Deer Lodge Valley, and to include the Blackfeet, Crows, Snakes, Bannocks, Flatheads, Pend-d'Oreilles, and Kootenays, and that the supplies for this superintendency be taken from St. Louis, by steamer, to Fort Benton. Four agents and a sub-agent would be required : an agent for the Blackfeet, one for the Crows, one for the Snakes and Bannocks, one for the Flatheads and Pend-d'Oreilles, and a sub agent for the Kootenays. This matter is well worthy the attention of the Indian Department, and to it I invite their attention, on the score of economy, efficiency, and security for the future.

POPULATION.

The white population, made up of Americans, French, and others, on and tributary to the road, may be estimated at ten thousand; found mostly at Walla-Walla and Lewiston, Deer Lodge, Hell's Gate, Beaver Head, Big Hole, Bitter Root, and Prickly Pear. They are engaged in farming, mining, transporting, merchandising, and the mechanic arts. As the mines develop, this population will be largely added to annually, both by emigrants by the way of the Missouri and from California. Already is activity infused into every branch of business, and the prospects for the future are flattering in the extreme. Two weekly papers are published in this region, at Walla-Walla the Statesman, and at Lewiston the Golden Age; schools, academies, and churches, already rear their heads, and the great number of books purchased and papers subscribed to bespeak the intelligence of the people. Every nationality, from John Chinaman to the Englishman, and every State in the Union, are here represented; but notwithstanding the variety of languages spoken and views entertained, harmony in council and uniformity in action as yet pervade and mark the body politic.

CATHOLIC MISSIONS.

The Jesuit Catholic fathers have three missions established along the line of the road; one among the Cœur-d'Alénes, one among the Pend-d'Oreilles and Flatheads, and one among the Blackfeet. The first site of the Cœur-d'Aléne mission was in the St. Joseph's Valley ; but the overflow of the stream and the many difficulties to which they were subjected at this point compelled them, in 1846, to abandon it in favor of its present site on the Cœur-d'Aléne River. They have erected here a fine church, dwellings, and such other buildings as are necessary for their wants ; the Indians are educated not only to worship God, but every attention is given to teach them to till the soil. The missions use Indian labor exclusively, under the direction of three lay-brothers, and are supported from a small fund for the "propagation of the Catholic faith," which is devoted exclusively to the purchase of those articles which they themselves cannot produce or make. They are thrifty and frugal, and by their zeal in the cause and devotion to the best interests of the Indians, for whom they have given their lives a voluntary offering, they wield an influence among the better portion such as no whites or government agents have ever been enabled to obtain. Far removed, as yet, from contact with civilization, their lives of upright moral rectitude, zeal in behalf of the Indian, their morning, noon, and night devotion, when all the tribes assemble and chant pæans to the Almighty, the perfect harmony that exists in their social family of Indians and half-breeds, has ever won my highest admiration. In all that tended towards the ultimate success of my movements, I have ever enjoyed their kind coöperation and zealous support, and during the many years spent near their mountain homes the kindest and warmest relations have ever existed between us.

The fathers, in abiding among the red men, have but one object in view : to rescue them from the blighting effects of ignorance and superstition, and to reclaim them from the effects of an advancing civilization, which to them is death. l can only regret that the results as yet obtained would not seem commensurate with the endeavors so manfully put forth. The only good, however, that I have ever seen effected among these people has been due to the exertions of these Catholic missionaries.

Many of these missions might be benefited by the government allowing them the charge of the schools and hospitals, for they actually take care of the Indians when sick and educate them when well, and all this with the mere pittance at their disposal, not a moiety of what they need ; while hundreds and thousands are squandered on paper for the benefit of the Indians, and which they never receive.

The Cœur-d'Aléne mission has the Fathers Josét and Gazzoli, and Brothers Francis, McGuine, and Campopiano. They have chosen a beautiful site, on a hill in the middle of the mission valley, and it has always proved to the weary traveler and destitute emigrant a St. Bernard in the Cœur-d'Aléne mountains. I fear that the location of our road, and the swarms of miners and emigrants that must pass here year after year, will so militate against the best interests of the mission that its present site will have to be changed or abandoned. This, for themselves and the Indians, is to be regretted ; but I can only regard it as the inevitable result of opening and settling the country. I have seen enough of Indians to convince me of this fact, that they can never exist in contact with the whites ; and their only salvation is to be removed far, far from their presence. But they have been removed so often that there seems now no place left for their further migration ; the waves of civilization have invaded their homes from both oceans, driving them year after year towards the Rocky Mountains; and now that we propose to invade these mountain solitudes, to wrest from them their hidden wealth, where under heavens can the Indians go? And may we not expect to see these people make one desperate struggle in the fastnesses of the Rocky Mountains for the maintenance of their last homes and the preservation of their lives. It is a matter that but too strongly commends itself to the early and considerate attention of the general government. The Indian is destined to disappear before the white man, and the only question is, how it may best be done, and his disappearance from our midst tempered with those elements calculated to produce to himself the least amount of suffering, and to us the least amount of cost.

The Pend-d'Oreille mission is pleasantly situated on Mission Creek, a few miles to the north of the Jocko reservation, where Fathers Minetry, Louis Vera Cruz, and Grassi preside, with two lay-brothers. They have here, as at the Cœur d'Aléne mission, a complete set of buildings for their residences and a beautiful church; here, too, Indian labor is employed. The original site of this mission was on the Clark's Fork, about fifty miles from Fort Colville, where the lower Pend-d'Oreille Indians lived; but, finding at this point more soil to cultivate and a better site, they removed hither in in 1855, bringing many of the lower Pend-d'Oreilles with them. They have, besides, a branch mission at Fort Colville, visited from time to time by Father Josét. They had a mission at the present site of Fort Owen till 1850, when it was abandoned and the property sold to Major John Owen, where he now resides, having a pleasant home, and the finest library I have seen on the north Pacific Coast.

The next mission is among the Blackfeet; but as yet not much headway has been made towards its permanent establishment. The fathers chose a site on Sun River, ten miles above the wagon-road crossing, where they erected a few buildings; but they never enclosed fields, and last year it was abandoned. They then held service at, and occupied the site of, Old Fort Campbell, a mile above Fort Benton. While I was at this point last summer they were projecting a site on the Marias with a view to there establish their permanent homes; the extent of good soil, its mild climate, its proximity to the homes of the Blackfeet, and its distance from the line of travel, all combining to determine them in its selection. At present the superior is Father Giorda, who has his headquarters at Fort Benton. The superior for many years while I was in the mountains was Father Congiato, from whom I have received many kindnesses and courtesies; but he being assigned to the presidency of one of the colleges at San Francisco, compelled a change. Fathers Imoda and Giorda, with two lay-brothers, are at the Blackfeet mission. The country and the Indians are mainly indebted to the zealous labor of the Reverend Father de Smet in establishing all these missions, for he truly is the great father of all Rocky Mountain missionaries. By his travels and his labors, and the dedication of his years to this noble task, he has left a name in the mountains revered by all who knew him, and a household god with every Indian who respects the black gown. His early work, called the " Oregon Mission," is replete with interesting information, and from which we ourselves have collected many geographical and statistical facts. To him and his colaborers in their self-sacrificing work I return my thanks for their many kindnesses, with the hope that they may live to see the full fruits of their zeal and toil amid the fastnesses and solitudes of the Rocky Mountains.

MINERAL WEALTH.

It has been only during the last three years that sanguine expectations have existed of mineral wealth to any great extent being found east of the Cascade Mountains or even in the northern portion of the Rocky Mountains proper. As soon as the Frazier River mines in British Columbia were discovered and the question of route by which to reach them in the shortest time and cheapest means was discussed, we found the eastern portion of Washington Territory ramified by hundreds of gold seekers in quest of this new Eldorado. This was in 1858. Gold was soon after discovered in the Wenatchee, Nachees, Okinagen, Simalkameen, and Clark's Fork, and worked till the Indians drove the miners from the country. In the succeeding year Captain Pierce, with a boldness and a judgment worthy every commendation, explored the Bitter Root Mountains, and the new discovery of the Nez Percés gold mines was the result. The wanderings of the miners in this region southward led to the Salmon River discoveries. But the gold miner, who is the most restless of mortals, did not rest content until he had crossed Snake River and discovered the Powder River mines; and not even then content, he opened up the Grand Ronde, Boisé, and Burnt River mines, in East Oregon, and elated at his success returns to his friends in West Oregon, taking the head of John Day's River in his route only to discover the richest quartz leads to be found outside of California.

His companions on Frazier's River heard of this new gold

field and they too must visit it, taking on their route the mouth of Clark's Fork and Spokane River, only there to discover gold which has since been taken out by the pound. This news spreads and the Powder River miner tracks it to the latter point, and thence to the Kootenay, where he is amply repaid for his toil and travel. While this is being done the Salmon River miner is not content with making twenty dollars per day, but he too must follow the Salmon River till it becomes a silvery thread in the mountains and crosses the range to discover the wealth of the Deer Lodge mines. As the weary emigrant crossing the plains hears of this Eldorado so near himself, he too must journey thither only to discover the rich gold mines in Big Hole and Beaver Head Valleys. Nor does it stop here, but the restless adventurer in St. Louis contracts the gold fever and threads the Missouri to Fort Benton in quest of the Deer Lodge mines, and while en route discovers the Prickly Pear gold mines. Thus, working like beavers, have the miners and emigrants crossed and recrossed the mountains during the last three years, ramifying in every direction until they have opened a gold region which, to-day, is sending to our mints a wealth equal to that of all California in her palmiest days.

Nor are these discoveries limited to the investigations of the miner or alone confined to the limits of our own territory, but the Catholic fathers in British Columbia have made equally rich discoveries at the headwaters of the Sascatchewan River and at Chief Mountain Lake, so that now gold is being profitably taken out at the following points : at John Day's, Grand Ronde, Powder River, Burnt, and Boisé, and Owyhee Rivers, in East Oregon ; in the Nez Percés mines, Salmon River, Spokane, Clark's Fork, Simalkameen, and

Deer Lodge, in East Washington Territory; and at Big Hole, Beaver Head, and Prickly Pear, in Montana.

The result of Captain Reynold's explorations would show that traces of gold were found by his party in all the tributaries to the Upper Yellowstone from the south. Enough discoveries have been made to warrant us in thinking that the entire mountain system will be found to be gold bearing.

Gold has also been discovered in the Big Blackfoot, Flathead, Kootenay, and Bitter Root Rivers, and slight traces in the Cœur d'Aléne and St. Regis Borgia.

I do not hesitate to say, upon the best of guidance, that the lower Clark's Fork will yet prove a rich gold region. The richest quartz leads yet found are in the John Day's. The other diggings are mostly surface or placer.

Wonderful have been the effects of this great alchemist in that quarter. It has transmuted sluggishness into activity, has brightened the dullest vision, elevated the industrious and frugal laborer, silenced the sceptic and caviller, and struck a new blow in behalf of a northern Pacific railroad route. The trade and travel along the Upper Columbia, where several steamers now ply between busy marts, of themselves attest what magical effects three years have wrought. Besides gold, lead for miles is found along the Kootenay. Red hematite iron ore, traces of copper and plumbago are found along the main Bitter Root. Cinnabar is said to exist along the Hell's Gate and at a point along the Upper Missouri. Coal is found along the Upper Missouri, and a deposit of cannel coal near the Three Buttes, northwest of Fort Benton, is said to exist. Near the headwaters of the Kootenay coal is also said to exist. Coal may yet be looked for east of the Cascades. Iron ore has been found near Thompson's Prairie on the Clark's Fork. Sul phur is found on the Lo-Lo Fork and on the tributaries to the Yellowstone; and a coal-oil spring is said to exist on the Big Horn River, a tributary to the Yellowstone.

It would be natural to suppose that the Rocky Mountain system, which, beginning far down in Mexico, there develops its wealth in gold and silver, and continuing northward to Pike's Peak would still retain its same geological character- istics as we trace them northward; and this we find to be the case, and on both slopes *at about the same altitude above the sea do we find these rich deposits.* What is true here is also true of the coast range, so that the Sierra Nevadas of California do not lose any of these characteristics when they become the Cascades of Oregon. We find that though for many years the gold region was limited to the western slopes and spurs of the Sierra Nevada, yet mineralogy pointed out that a Washoe *must* exist, and hence develop- ments of fabulous wealth were made on the east slopes of the Sierra Nevada, only to be followed by equally rich de- posits now being worked on the eastern slopes of the Cas- cades, in Oregon.

These great mineral deposits must have an ultimate bear- ing upon the location of the Pacific railroad, adding, as they will, trade, travel, and wealth to its every mile when built.

BUILDING MATERIAL.

The great depots for building material, stone, timber, and limestone, exist principally in the mountain sections, but the plains on either side are not destitute in this particular; white and red fir, and white and yellow pine are found at Walla-Walla, in the Blue Mountains, on the Clearwater where cedar is added, and all through the Bitter Root and Rocky Mountains the finest white and red cedar, white pine and red fir, that I have ever seen are found.

Between the Cœur-d'Aléne mission and Hell's Gate, spruce, hemlock and birch are also found. The elevation at the Dalles and the line of longitude passing through it is the eastern limit of oak, which is traced from the coast eastward. We have no hard woods in this region, and, indeed, nearly all the hard woods used in California and Oregon are shipped around Cape Horn. Even the oak found in the Pacific is of an inferior growth, generally wind shaken and but little used in the mechanical arts. Timber is found until within 120 miles of Fort Benton ; it grows also along the Upper Missouri and all its tributaries where it can be rafted down. All the rivers here lend themselves admirably to rafting purposes.

Limestone is found on the Clearwater, near the Cœur-d'Aléne Lake, in the Cœur-d'Aléne Mountains, in Hell's Gate, in the Bitter Root Valley, and head of the north fork of the Little Blackfoot River. Beautiful red sandstone is found along the Bitter Root River. Slates are found along the Cœur-d'Aléne, St. Regis Borgia and Bitter Root. Excellent gray sandstone is found in the Bitter Root Mountains, from which grindstones and scythe-stones have been made, and good burr-stones for mills have been got out of the mountains and used by the Jesuit fathers. Good sand is found along the Columbia, along the Spokane and Cœur-d'Aléne and along the Hell's Gate Rivers.

AGRICULTURAL LANDS.

The amount of agricultural land along the general line of the road may be safely estimated at one thousand square miles, or six hundred and forty thousand acres.

The largest single body is found in the Walla-Walla Valley, where its rich soil, freedom from early frosts, with the mild climate there found, constitutes it a great agricultural centre. Corn, oats, barley, spring and autumn wheat, tobacco, and every variety of vegetable, are here grown in great abundance; wheat, thirty bushels to the acre; barley and oats, forty bushels, and corn from sixty to eighty bushels, and potatoes from three hundred to six hundred bushels to the acre. The grape and peach will be found to grow well here, and when the country is settled the oak will take its place among the chief forest growth.

The mines furnish a ready market for all that is raised, cash payments for which are made in gold.

Fruit-growing is attracting attention and thus far promises well. Erection of grist-mills keep pace with the growing demands of the country. Another favorite agricultural tract is the Dry Creek; its valley is already studded with beautiful farms; so also the Touchet, where there is still room for the industrious emigrant.

At the mouth of the Palouse is a small tract of good land, enough for one small farm, and which, as trade and travel increase, must become an important point.

Several tracts of good soil are found along the Palouse, but the absence of timber is an impediment not easily supplied.

Small tracts are found along the Cow Creek and along the

road to Aspen Grove, and many in the direction of the long line of lagoons found along the road.

An excellent body of good land is found along Lake Williamson, and along several of the smaller lakes and creeks to its mouth.

Several small bodies are found along both banks of the Spokane ; near the Foot Hill several farms are already under culture. Along the Upper Spokane are also small bodies. At the Cœur-d'Aléne mission is a body of five or six square miles of most beautiful soil ; several hundred acres are here under cultivation by the mission and the Indians. Oats, barley, wheat, peas, and potatoes, are raised in rich abundance.

One of the largest bodies of good land is in the valleys of the St. Joseph's and Cœur-d'Aléne, and if these valleys are once drained, a body of forty thousand acres of the finest soil in the world will be reclaimed—soil six and eight feet deep and as black as a coal. This overflow can be prevented by widening the natural outlet, and making an artificial one along side of it ; an appropriation of five thousand dollars will meet this difficulty, and the ends to which it looks are well worthy the experiment. Rock blasting is the only method of accomplishing it, and should be done during a low stage of water. The overflow is alone occasioned in the highest stages of water, when the mouth of the outlet of the lake is not capacious enough to discharge the volume of water sent into it by its feeders.

On the right bank of the Cœur-d'Aléne, opposite the mission, is a mile square of beautiful soil. Four-Mile Prairie has a good body of land, also Ten-Mile Prairie.

From this point to the Bitter Root Ferry, I fear the frosts

2*

and other mountain characteristics will preclude any farms from being opened.

At the mouth of the St. Regis Borgia, however, several tracts of land are found, which, if cultivated, will come into requisition for mail stations, and supplying travelers and emigrants. Seven-Mile Prairie along the Bitter Root River offers a good site for small farms; also Brown's Prairie, Nemoté Prairie, Skiotay Creek, where the wild timothy abundantly grows, and many large and beautiful tracts along the right bank of the main Bitter Root. Frenchtown, in Hell's Gate Ronde, already contains from ten to fifteen small farms, and there is room for many more. The small creeks in Hell's Gate Ronde offer many choice sites for farms; a dozen at least are here now under cultivation. Wheat, potatoes, oats, and barley, and all vegetables are raised. Corn is not matured, but is raised for roasting ears. The St. Mary's Valley to the south, and the Jocko and Flathead Valleys to its north, are large and favorite farming sections.

Among the small creeks tributaries to the Hell's Gate, four and five miles from Big Blackfoot, at Clark's camp, at Beaver Dam Butte, and along some of the smaller water-runs good soil is found. The upper portion of the valley of Flint Creek may be found suited to agriculture.

From Flint Creek to the American Fork are many beautiful localities where farms may be opened. A large and beautiful one is already opened by a Mr. Dempsey, where he grows oats and wheat luxuriantly. The American Fork has some beautiful bodies of land along it. The Deer Lodge is a large and well located valley and contains from fifty to one hundred and fifty square miles, where crops have already been grown. Along the Little Blackfoot and its smaller tributaries are found patches of good soil. From

this point to the Big Prickly Pear Creek the altitude and
wet condition of the soil may prove impediments to successful or economical culture.

Crossing the divide we get into a new climate and change
of soil ; along the Big Prickly Pear will be found several
small and choice localities for farms, and if the mines on
the eastern slope prove successful, I look forward with much
hope to see all these creeks settled and fine farms developed
under the hand of the Rocky Mountain farmer. All the
small creeks from the Big Prickly Pear to the Bird Tail
Rock have smaller bodies of good soil, and if the mines
prove a reality a market will be at hand for all they can
raise.

To the south of the Bird Tail Rock, about three to five
miles in a wild broken region, through which a wagon road
can be laid to shorten our line from Sun River to the Dearborn, is one of the largest and richest bodies of land that I
have seen east of the mountains, and where timber from the
mountains and rock at hand will supply all the requisites
of the first settlers. Situated, as it is, midway between
Fort Benton and the Prickly Pear gold mines, it is worthy
an examination on the part of those who propose to settle
in the country. I examined it on my return from Fort
Benton, June, 1862, to see if we could not shorten our line
by going south of the Bird Tail Rock. With moderate work
I found this feasible, and saving from eight to ten miles in
distance ; and it was on this trip that I was struck with the
richness of the soil and the extent of agricultural land here
found. It is a favorite resort for game, and sheltered, as it
is, by high rock walls from the cold bleak winds which sweep
across the plains, forms a choice site for farms.

The next good tract is along the Sun River, where a stiff

clay soil is already cultivated to advantage. The Blackfoot agency has a farm of from thirty to forty acres under cultivation, and raises wheat and vegetables. Its valley is from fifty to one hundred and fifty feet below the general level of the Missouri plains, and is thus sheltered from the colder winds that sweep across them at spring seasons. Some patches of good soil are found along the Missouri above Fort Benton, but their culture has never been tested. There are enough statistics of growth from actual cultivation along the line to show that the many acres herein referred to will grow abundant crops, and all of which will come under tillage as soon as a market presents itself. At all the points referred to abundant building material and water are at hand, and mill sites can be had on nearly every water-course. There are already three grist and two saw mills in the Walla-Walla Valley, one grist mill at the Cœur-d'Aléne mission, one saw and one grist mill at Frenchtown, one saw and one grist mill at the Pend-d'Oreille mission, one saw mill at the Jocko River, and one saw and one grist mill at Fort Owen; one steam saw and one grist mill owned by La Barge & Co., which they propose to erect near the Deer Lodge gold mines. The Blackfoot agency use a small hand-mill for grinding their wheat.

As an experiment, the value of which another season will develop, I had shipped from St. Louis to Fort Benton, and thence packed over into the Bitter Root Mountains, twenty-four bushels of blue grass seed and twelve bushels of clover seed, and sowed them at the crossing of the Bitter Root, along the valley of the St. Regis Borgia and the Cœur-d'Aléne to the mission, wherever the grass was sparse. The seed was thrown broadcast over the ground and through the woods, and over the prairie, at such points as are likely to be selected as camping-grounds. This wooded section of one hundred miles is the only point where grass is at all scarce, and this experiment was made by myself to meet fully all the wants of a future and large emigration over the road.

GRAZING CAPABILITIES.

The experience of all persons traveling through this region has been that, from the Columbia to the Missouri Rivers, finer grasses have never anywhere been seen; the number and condition of the stock that feed upon the wild grass alone shows both abundance and nutrition. Wild hay can be, and is, cut from thousands of acres. The grass is mostly a wild bunch grass, growing from twelve to eighteen inches high, and covering the entire country. Horses and horned-stock by thousands, and sheep by hundreds, all bespeak the wealth that is wrapped up in the native grasses of the north Pacific region, and I confidently look forward to seeing the wealth of these beautiful mountain valleys yet consist in the thousands of fleecy flocks to be here sheared ; and if the streams of the Rocky Mountains are caught and harnessed to the spindles and looms of wool manufactories to be there erected, that the annual shipments of wool to a St. Louis market will constitute a trade replete with wealth and magnitude. I believe it to be an ordination of Providence that as the buffalo that now blacken the western plains by their millions of shaggy coats disappear with the red man, whose sustenance they now are, their place will be supplied by the silvery fleeces of millions of sheep tended by white men, which this region is capable of sustaining.

The wolves now exist in places as an impediment to much attention being given the subject, but travel and settlement soon destroy them. The premiums on cotton which is soon to be transferred to wool render this question one of great magnitude for this region, and I yet hope to see the genial valleys of the Rocky Mountains send to a St. Louis market, *via* the Upper Missouri River, its cargoes of wool and its shipments of gold, as its natural tribute to our national wealth.

CLIMATE, SNOW AND COLD.

There has been no one subject so little understood or so much misrepresented as the climate of the northern valleys of the Rocky Mountains and the plains extending to their either base. I am frank to admit that the section of our road from the Cœur-d'Aléne mission to the Bitter Root Ferry does interpose the obstruction of snow to such an extent that I despair of seeing it traveled in winter unless a daily mail coach is placed upon the line, when the snow being beaten down twice a day, would, I think, keep the line constantly open. But all the remaining sections are mild, with so little snow that traveling with horses can be kept up all winter. And although the climate in the region first referred to is severe, by going north to the Clark's Fork, we at once enter a milder section, and one that offers every advantage to travel. The temperature of Walla-Walla, in 46°, is similar to that of Washington City, in 38° latitude; that of the Clark's Fork, in 48°, to that of St. Joseph's, Missouri, in latitude 41°; that of the Bitter Root Valley, in 46°, is similar to that of Philadelphia, in latitude 40°, with about the same amount of snow, and with the exception of a few days of intense cold, about the same average temperature. This condition of facts is not accidental, but arises from the truths of meteorological laws that are as unvarying as they are wonderful and useful. As early as the winter of 1853, which I spent in these mountains, my attention was called to the mild open region lying between the Deer Lodge Valley and Fort Laramie, where the buffalo roamed in millions through the winter, and which during that season constituted the great hunting grounds of the Crows, Blackfeet, and other mountain tribes. Upon investigating the peculiarities of the country, I learned from the Indians, and afterwards confirmed by my own explorations, the fact of the existence of an infinite number of hot springs at the headwaters of the Missouri, Columbia, and Yellowstone Rivers, and that hot geysers, similar to those of California, existed at the head of the Yellowstone; that this line of hot springs was traced to the Big Horn, where a coal-oil spring, similar in all respects to those worked in west Pennsylvania and Ohio, exist, and where I am sanguine in believing that the whole country is underlaid with immense coal fields. Here, then, was a feature sufficient to create great modifications of climate, not local in its effect, but which even extends for several hundred miles from the Red Buttes, on the Platte, to the plains of the Columbia. The meteorological statistics collected during a great number of years have enabled us to trace an isochimenal line across the continent, from St. Joseph's, Missouri, to the Pacific; and the direction taken by this line is wonderful and worthy the most important attention in all future legislation that looks towards the travel and settlement of this country. This line, which leaves St. Joseph's in latitude 40°, follows the general line of the Platte to Fort Laramie, where, from newly introduced causes, it tends north-westwardly, between the Wind River chain and the Black Hills, crossing the summit of the Rocky Mountains in latitude 47°; showing that in the interval from St. Joseph's it had gained six degrees of latitude. Tracing it still further westward it goes as high as 48°, and develops itself in a fan-like shape in the plains of the Columbia. From Fort Laramie to the Clark's Fork, I call this an atmospheric river of heat, varying in width from one to one hundred miles. On its either side, north and south, are walls of cold air, and which are so clearly perceptible, that you always detect when you are upon its shores.

It would seem natural that the large volume of air in motion between the Wind River chain and the Black Hills must receive a certain amount of heat as it passes over the line of hot boiling springs here found, which, added to the great heat evolved from the large volumes of water here existing, which is constantly cumulative, must all tend to modify its temperature to the extent that the thermometer detects. The prevalent direction of the winds, the physical face of the country, its altitude, and the large volume of water, all, doubtless, enter to create this modification; but from whatsoever cause it arises, it exists as a fact that must for all time enter as an element worthy of every attention in lines of travel and communication from the eastern plains to the north Pacific. A comparison of the altitude of the South Pass, with the country on its every side, with Mullan's Pass, further to the north, may be useful in this connexion. The South Pass has an altitude of seven thousand four hundred and eighty-nine feet above the level of the sea. The Wind River chain, to its north, rises till it attains, at Frémont's Peak, an elevation of thirteen thousand five hundred and seventy feet, while to the north the mountains increase in altitude till they attain, at Long's Peak, an elevation of fifteen thousand feet ; while the plains to the east have an elevation of six thousand feet, and the mountains to the west, forming the east rim of the great basin, have an elevation of eight thousand two hundred and thirty-four feet, and the country between it and the South Pass an elevation of six thousand and two hundred and thirty-four feet above the level of the sea. The highest point on the road in the summit line at Mullan's Pass has an elevation of six thousand feet, which is lower by fourteen hundred and eighty-nine feet than the South Pass, and allowing what we find to be here the case, viz. : two hundred and eighty feet of altitude for each degree of temperature, we see that Mullan's Pass enjoys six degrees of milder temperature, due to this difference of altitude alone. At the South Pass are many high snow peaks, as Frémont's Peak, Three Tetons, Laramie Peak, Long's Peak, and others, all of which must tend to modify the temperature ; whereas, to the north we have no high snow peaks, but the mountains have a general elevation of from seven to eight thousand feet above the level of the sea, and of most marked uniformity in point of altitude.

The high range of the Wind River chain stands as a curvilinear wall to deflect and direct the currents of the atmosphere as they sweep across the continent. (By-the-bye, whence arises the name of the Wind River chain ?) All their slopes are well located to reflect back the direct rays of the heat of the sun to the valleys that lay at their bases. These valleys, already warm by virtue of the hot springs existing among them, receive this accumulative heat, which, driven by the new currents of cold air from the plains, rises and moves onward in the form of a river towards the val-

leys of the Rocky Mountains, where it joins the milder cur-
rent from the Pacific and diffuses over the whole region a
mild, healthy, invigorating, and useful climate.

RIVERS AND WATER-COURSES.

The entire length of road from the Columbia to the Missouri may be said to follow one continuous line of water-
courses, one of the chief recommendations of which is that
not one is alkaline ; but the water is of a delicious purity.

This fact on a long line of march with much stock is an
advantage that all who have crossed the plains will not fail
to appreciate.　In the first stretch of fifty miles we cross the
Walla-Walla, Mill Creek, Dry Creek, and Touchet River.
They all rise in the Blue Mountains, the last three flowing
into the Walla-Walla River, and the last into the Columbia, at the town of Wallula.

The Snake River, or the south fork of the Columbia, s
the most considerable stream crossed ; it rises in the Wind
River Mountains, and is about one thousand miles long.　It
is navigated by steamers of heavy draught as far as Lewiston till the first of August, and light draught steamers have
run as late as the 29th of September ; and, during the coming year, it is proposed to run steamers as far as Fort Boisé,
to supply the mines on either side at this point.　Steamers
have run up the Clearwater above the Lapway, but only
during the highest stage of water.

There is no timber along this portion of the Snake River,
and steamers have to depend on drift wood, of which there
has always been ample up to date.　But fuel can be rafted
down from the Clearwater.

The importance of the Snake River for purposes of steam navigation begins now to be thoroughly appreciated. The cost of transportation is a large item, but as long as the mines continue to pay as well as they now do, so long will it pay to run steamers on this river. The scale upon which nature has worked in this region is wonderful. Though she has interposed great physical obstacles, both in rivers and on land, to the opening of the country and speedy travel, she has, nevertheless, at the same time placed in the same region great areas of gold fields, the wealth from which is destined to so meet these obstacles that it is only a question of time when this region will boast of as rapid and as cheap travel as regions where the physical difficulties have been fewer, but the ends sought not so stimulating as the search and mining for gold has proved to be.

The next stream of note is the Palouse, which, rising in the Bitter Root Mountains, has a length of ninety miles, unsuited to either navigation or rafting purposes. Its banks are not timbered for a stretch of forty miles from its mouth. Its minor tributaries, north and eastward, only enter as elements of advantage for camping purposes. The great number of lakes found on the plains of the Columbia enter as marked features in this immense area.

Hangman's Creek rises in the Bitter Root Mountains, has a length of fifty miles, and empties into the Spokane; has no special value outside of affording camping facilities. The Spokane is the outlet of the Cœur-d'Aléne Lake, is one hundred miles long, and drains throughout a timbered region, and enjoys special advantages for rafting purposes. Its mouth is the southern limit for timber along the main Columbia.

Gold mines exist at and above its mouth. A project has

been conceived to open a canal from the Clark's Fork at the
Pend-d'Oreille Lake to the Spokane, the elevation being
suited for this purpose, and the Clark's Fork, with the Pend-
d'Oreille Lake, offers marked advantages for steam navigation for at least eighty miles. My idea has always been
that steam on the Clark's Fork, at Park's crossing, running
thirty miles to the lake, and thirty miles across the lake, and
twenty miles up the Clark's Fork, would reach a point
where a cheap wagon road could be opened to the Hell's
Gate, and that for all purposes of travel, carrying mails, and
emigrant purposes, that in time this would be done.

I do not know what special advantages are expected to be
derived from the canal referred to, nor do I think it will be
undertaken unless this region should prove a very rich and
extensive gold field. The Cœur-d'Aléne Lake and its two
arms can be navigated by steam, and enjoy special advan-
takes for rafting purposes.

The Clark's Fork, which is the middle fork of the Columbia, is a most important stream, and enjoys preëminent advantages for rafting purposes. It is fed by two large forks,
the Flathead River from the north, and Bitter Root from
the south.

The St. Regis Borgia enjoys no special advantages outside
of defining a valley, and being useful for camping purposes.
Its frequent crossings have rendered it rather an element of
disadvantage.

The Bitter Root River, with its three large tributaries,
Big Blackfoot, Hell's Gate, and St. Mary's River, enjoys
special advantages for rafting purposes, and though here and
there are stretches suited to steam navigation, yet, broken
as it is in its length by fall and rapids, precludes the possibility of its ever being used for such purposes.

Crossing the Rocky Mountains we find none of the tributaries of the Upper Missouri along our immediate line enjoying any special advantage, except the Dearborn and Sun Rivers, both of which might be used for rafting timber and stone from their headwaters; the Sun River more so than the Dearborn. The Upper Missouri, with its three forks, enjoys special advantages both for navigation and rafting purposes. The distance from the falls of the Missouri to the Three Forks is about one hundred and seventy-five miles, and though the water is rapid at a few points, yet in the lowest stage of water steamers can run; and here is a field for enterprise and bold experiments never as yet tested, and which cannot long lie dormant if the Prickly Pear, Deer Lodge, and Beaver Head mines prove a growing reality.

At present steamers from St. Louis land at Fort Benton, but they can run forty miles further to the foot of the falls. If steamers are put on above the fall a portage of fifteen miles becomes necessary.

The mouth of a small creek from the south, called Portage Creek, affords a good landing site. The river from this point to Fort Benton is broad and bold, with plenty of cottonwood fuel on the small creeks and on the main river.

The portage will have to be made on the south bank, as the great number of coulées and broken country on the north bank will render a road expensive, if not impracticable. On the south bank, by keeping back, say two or three miles from the river, we head all the coulées, and have a level and beautiful plain till reaching the head of the falls, when you descend to the river bank by a gradual slope at the White Bear Islands, just below the mouth of Sun River, where the Missouri is again a clear, bold, and placid stream.

The great falls of the Missouri form the only feature of

great beauty in this region, and are well worthy a visit. The falls begin nine miles below the mouth of Sun River, and continue in a stretch of thirteen miles with rapids, falls, and cascades, gaining in this distance a total fall of 380 feet.

The lower falls are the largest, being 84 feet, where the stream is 480 yards wide, the southern half of which, in a single sheet, leaps over a ledge of sand-stone rocks, while the northern half, in a series of fall, cascades, and chutes, makes a broken descent into the basin lying at its foot. The whole rock formation here is a yellow sandstone, and the character given to the falls has depended upon the ease or difficulty with which the water has either washed the rock away to a sharp face, as in the southern half, or worn it away into plateaus and benches, as in the northern half. A high, prominent point projects into the river at the foot of the falls from the northern Bank, from which you enjoy a beautiful view.

The water, the rock thinly veiled in foam, the high, bold banks, the wildness of the scene, so far from civilization, the thundering noise of the fall of the cataract, the rising of the mist, the *tout ensemble*, constituted a picture worthy the pencil of the artist and the toil of the tourist. The upper, but smaller, falls are 40 feet high, but very beautiful. Between the upper and lower are the Horse Shoe and other falls, from three to fifteen feet.

The banks on either side continue to retain throughout this entire length their wild, rocky, rugged character, so that the Missouri itself ploughs in a deep, narrow gorge one thousand feet below the general line of the plains that bound it on either side.

The special advantages enjoyed by the Columbia and Missouri Rivers, rising, as they do, in the same sources in the

Rocky Mountains, and flowing thousands of miles into two oceans, must ever enter as essential and economical elements into trade, travel, and railroad construction across the continent.

The Columbia is the only stream that rises in the main chain of the Rocky Mountains, breaking through the coast range and emptying into the Pacific. The large watercourses, with their great net-work of tributaries, lying between the 44th and 49th degrees of north latitude, must ever cause this region to enjoy facilities given by nature to no other west of the Mississippi. These large bodies of water tend to modify the climate, supply mill sites, water farms, and grazing fields; enable the miner to work miles of sluice-boxes, the merchant to float, by steam, his wares to the very heart of the Rocky range, and stand ever ready and panting to be converted into steam for the iron horse that must soon invade their dominions.

MOUNTAIN SYSTEM.

The early geographers who attempted the mapping of the country west of the Mississippi left us a very vague and erroneous outline of the Rocky Mountain formation, or the direction of the vast system of spurs that go to form them. The dividing ridge of the Rocky range was nearly always represented as a right line trending from northwest to southeast from our northern boundary to New Mexico. This right line has, however, disappeared from our maps, as explorations have brought in from year to year the results of their researches.

The greatest deviation from a right line that occurs in

close proximity to our road exists from the head of Deer Lodge Valley to the Wind River chain, where the mountains enclose in a large reëntering angle the Big Hole Valley. I look forward to seeing this section become a great centre both for mining and agricultural developments. Our barometer detects this fact, that all the waters of the Clark's Fork, with the Bitter Root, Hell's Gate, St. Mary's, and Big Blackfoot, flow in a basin elevated above the waters of the plains of the Columbia by seven hundred feet, and above the plains of the Missouri by five hundred feet.

The waters of the south fork of the Columbia, rising as far south as the 40th parallel of latitude, all flow northward; so, also, the Missouri and the Yellowstone, showing that they rise in higher mountains, and necessarily the points to which they flow must be lower, and hence, as you go northward within certain limits, the country must be lower, and the mountain passes have a less altitude above the sea level. Our barometer confirms this theory, and hence, from the 42d to the 48th parallel, you will find the lowest passes in the Rocky range.

The mountain formation, with its great system of spurs, can only be likened to the coral formation, as difficult to delineate on the maps, and following a rule as difficult to describe.

PRESENT WANTS OF THE COUNTRY.

The results of our long explorations in the country have developed the following results, to which the attention of Congress should be especially invited, and which, if met in the liberal spirit they so well merit, will redound to the best interests of the government by testing, to the fullest extent, the golden resources of the Rocky Mountain region, and which, standing, as they do, midway between the Mississippi and Pacific, will develop a bond of union and strength between the extreme sections of the continent.

These are to establish a military depot at the head of the great falls of the Missouri, at the mouth of Sun River, with a three-company cavalry post; this depot to supply a four-company cavalry post at the Deer Lodge, and a one-company post at Hell's Gate. In this region are near fifteen thousand Indians, and these posts are needed to guard the emigrant lines that here fork, and over which travel will pass every season, as well as to protect the hundreds of miners and farmers there found. The supplies for all these points should be shipped from St. Louis by steamres direct to the falls of Missouri. These posts can be economically maintained and are a military necessity for the country. The military post at Walla-Walla should be reduced to a depot and its troops moved to Fort Boisé. One company should be posted at Florence, on the Salmon River; one at Oro Fino, and one company either at the Cœur-d'Aléne mission or at Park's Crossing, on the Clark's Fork. Four companies should be stationed at Fort Hall and the military post at Fort Laramie retained with a large garrison. These are all needed for the protection of overland emigrants. The supplies for all these points, except the latter, should come from Walla-Walla. Military roads should be opened from Deer Lodge to Salt Lake, from Fort Benton to Beaver Head Valley, and from Beaver Head Valley to Salmon River or to Florence.

The entire mountain region, from the Salmon River mountains to the plains at the head of the Platte, should be

3

explored, mapped, and reported upon, and a short and di
rect connexion opened between Fort Laramie and the Salmon River gold mines, as it may be found that the shortest
and best route for a Pacific railroad to the Columbia may
be found from the Platte, *via* Salmon River, to Walla-
Walla; at any rate, it is well worthy a special examination.
The government should also test the capabilities of the Upper Missouri and Yellowstone for steam navigation both for
national purposes and as aids to the building of the Pacific
railroad.

The Indians should be concentrated as much as possible
and a Rocky Mountain superintendency established, with its
headquarters either in Deer Lodge or in Beaver Head Valley, and should include the Crows, Blackfeet, Snakes, Bon-
nacks, Flatheads, Pend-d'Oreilles, Kootenays, and the Mountain Nez Percés.

A Columbia River superintendency should be established,
with its headquarters at Walla-Walla, and to include all the
Indians between the Rocky and Cascade range, from the
49th parallel to Utah, both in Oregon and Washington.

Mail facilities should be established from Hell's Gate to
Fort Benton, and from Deer Lodge to Fort Laramie and
Salt Lake, also from Beaver Head Valley to Florence City.
A new military department of Oregon and Washington
should be established, with its headquarters at Walla-Walla ;
a first-class military road should be opened from Wallula
to Puget Sound; Portland, Puget Sound, and Salt Lake
should be connected by a military telegraph; a new territory should be established east of the Cascade Mountains, and
a detailed map of its topography at once made by the General Land Office. A branch mint should be put into operation at once either at Portland or east of the Cascade Mts.

GEOGRAPHY, TOPOGRAPHY AND RESOURCES

OF

THE NORTHWESTERN TERRITORIES.*

IN compliance with your polite invitation to lay before
the Society such new geographical facts as my explorations
in the Rocky Mountains have developed, I desire to make
in brief outline a few references to the early history of a region replete with wild romance, from the date of its first
exploration, when the red man reigned alone in its solitudes
and its wilds, to the present period, when it has become the
home of your friends and brothers, and men that are laying
the foundation of a mighty empire, adding strength and
wealth to our national structure.

When the existence of the western world was demonstra-
ted to the mind of Columbus, it was not astonishing to know
that his highly-wrought imagination revelled in pictures of
golden wealth, which the survey and exploration of this
world would develop; though even at this day we are astonished to know, that for a period of nineteen years he was allowed to importune his government for permission to verify

---

* An Address delivered by Capt. Mullan before the American Geographical and Statistical Society of New York.

his own firm, unalterable convictions that great Nature had here carved a new continent, where man, in time, was to solve a new problem in human destiny.

Though the American continent was discovered in 1492, three hundred years were allowed to elapse before the geography of its western half attracted attention, either in public councils or private circles, and it was for a period of thirteen years more that endeavors were made and set on foot that should prove to the world the importance of that vast area which to-day we are carving into new States, the power of which is fast revolutionizing the direction of the commerce of the world. English, Spanish and Russian navigators had coasted along the shores of the North Pacific as early as the sixteenth century, but, with the exception of marking in a very general and incorrect manner the coast line from Mexico to the Russian possessions, no special discoveries were made. No navigator had pointed out the existence of our majestic Columbia, nor could give us an idea of the character of the interior of half our continent.

A mythical river, called San Roque, it is true, had existed in the wild imagination of Jonathan Carver many years before our great Pacific feeder was discovered, but nothing sufficiently authentic was believed that warranted the geographical world to put forth measures to confirm the truth of the statement. The importance of the North Pacific to the geographical world, therefore, may be said to date from the discovery of the Columbia River, but to the American people it became of marked interest only from the date of the Louisiana purchase.

Before Mr. Jefferson was accredited to the government at Paris as our Minister Resident in 1792, his attention had been called to the geography of the North Pacific, and it was his own philosophical mind that pointed out the necessity of the existence of some such river as the Columbia.

He reasoned thus: That where the eastern water-shed of the Snowy Mountains, as the present Rocky Range was then called, poured the meltings of its snows in such a reservoir as the Missouri, there must be a corresponding shedding of waters on the western slope that must feed some large reservoir where the Columbia has since been discovered to exist. While representing our government in France, he lost none of his interest in this, to him, most interesting subject, but he carried it to the extent of inducing Ledyard, the the famous traveller, to change the scene and locale of his labors, and leave the Old for the New Continent, to link his name to a field of exploration destined to immortalize any man that would initiate it.

The enthusiastic mind of Ledyard grasped the gigantic idea, and through letters from Mr. Jefferson was duly accredited to the Empress Catherine, then upon the throne of Russia, reached St. Petersburg, and started through Western Russia for the Pacific, intending to come down along the western slope of the continent until he should discover the mouth of this philosophical river, the existence of which no navigator had discovered, and of whose wealth no explorer had dreamed until our own day and generation. The necessary passports were granted by the Empress, and Ledyard was in full progress to fulfil his glorious mission, when the Empress recalled her authority, had Ledyard followed, overtaken and ordered back to Paris, on the ground of being a spy in the joint interests of the French and American Governments.

The short vision of the Russian Government was not allowed thus to interpose itself as a barrier to the full development of this philosophical problem. Though Ledyard was forced to give up his much-coveted mission and return to the exploration of the Nile, the waters of which finally gave him a burial, yet Jefferson determined to pursue this project with unabated zeal and with new incentive, for now was formed for the first time the determination on the part of the National Government—through Mr. Jefferson—to purchase the territory of Louisiana, which, however, was not effected until 1803.

The Columbia River was in the interval (1798) discovered by a Boston shipmaster, Grey, who, in honor of the first ship that its majestic waters bore, named it the Columbia. When this discovery was made and an American purchase had been effected, an American exploration and occupancy was determined upon. The purchase of the territory of Louisiana, which the entire country, (excepting the Mexican domain,) from the mouth of the Mississippi to its sources and westwardly to the Pacific, was called, was effected not only in demonstration of the principle of territorial expansion within the limits of the North American Continent, but in vindication of the determination that the Americans as a people alone are to occupy and govern the major portion of the North American Continent, untrammelled by foreign friends and unmolested by domestic foes. Though Mr. Jefferson had some scruples as to his right to make the purchase under the authority given to Congress in the Constitution, these scruples were all passed over and lost sight of in the importance and magnitude of the legacy he desired to bequeath his country. Even before the purchase was complete and the French flag had given place to our own national colors, the project was set on foot to explore and make known, not only to us but to the geographical world, the character and importance of this vast area, the solitude of which had never yet been invaded by either the tramp of the pale face, or his invigorating influence.

This project of exploration was conceived, initiated, its details supervised and put into execution by Mr. Jefferson himself, who, as President of the United States in 1804, called Captains Lewis and Clark to its exploration, and gave them unlimited authority to open and reveal the contents of this new book of geographical and physical wonders.

From this exploration, which occupied a period of three years, dates, therefore, the time when it attracted an additional attention from the American public. Lewis and Clark, starting from the then small French village of St. Louis, traced the Missouri to its sources in the Rocky Mountains, crossed the range by a practicable though difficult pass, reached the headwaters of the Columbia, and followed this noble stream to its junction with the ocean. Though much was done for the geographical world by this exploration, it was left for future years to point to this region as one where we should build up an element of strength in our national progress. Though the record of these celebrated travellers found many readers and admirers, it had not the effect to attract to the shores of the Pacific a colonial outpost, for it was not till a much later date that an American occupation was had: not, indeed, until the subject of the American right to the region was being discussed in our legislative halls, with a fair prospect of a war with England being upon our hands, did we find ourselves compelled to assert our right, to maintain our supremacy, and people with Americans this our Pacific domain.

It was not till 1834 that we found the western frontier emigrant determined to take up his abode on the shores of the western sea, and evince a willingness to wend his way over sterile plains and miles of arid wastes, the pioneer of thousands who have since followed in his wake. But so few had taken up their abode as late as 1847, that Congress, both in its liberality and in the spirit of inducement to form an American colony, offered munificent bounties of land to those who would take up their abode in Oregon. It was not, therefore, until the conclusion of the Mexican war that, as a nation, we appreciated that importance of our Pacific possessions, which their location then, and wealth since, have so fully entitled them to. In 1837, a territorial organization was given to Oregon, and on the conclusion of the Mexican war we gave to California a military government, preparatory to affording her the civil machinery which, during a period of fourteen years, has worked smoothly and efficiently for her own interests, and the nation's advancement.

The exploration of Fremont that had preceded our difficulties with Mexico, all tended to show that the great chain of the Rocky and Sierra Nevada Mountains interposed themselves as great physical barriers to such an extent, that a journey across the continent was only to be made at the risk of comfort, privation, difficulty, if not life itself; and it was natural to suppose that when the first gold discoveries were made, attracting to their workings thousands of hardy miners, that some route fraught with fewer discomforts and difficulties would be sought, by which its population could be largely added to, and by which their various wants would be supplied.

The route via Cape Horn was, of course, the first thought of and the first travelled, but the length of the journey and the loss of time all determined upon the nearer and better route of the Isthmus of Panama and Nicaragua; and for four years did a wave of population, concentrating from all quarters of the globe, flow over this narrow neck of land, *en route* to that new El Dorado, the importance of which is marvellous, and which increases as time advances. But though these lines were had and travelled, our wants were far from being subserved; and the question how to reach them in the safest, best and shortest time was discussed, until the discussion eventuated in a resolve on the part of the General Government to explore the entire continent from the Mississippi to the Pacific, to discover a practicable route for a railroad line to supply this desideratum. From this date my own connection with the exploration of our Northwest interior, which, continued as it has been for the greater portion of the last ten years, has familiarized me somewhat with the geography of the Northwest Pacific, and which, in no spirit of disparagement to other sections, is, in my judgment, destined to occupy an important place in the growing interests of our nation. The law of Congress that called several corps of engineers and explorers to examine the country for a practicable highway to the Pacific, was executed in 1853, and to one of these fields of exploration was assigned the late General Isaac I. Stevens. This field extended from St. Paul's, Minnesota, to the Columbia River, thence northward to Puget Sound. It was subdivided into two main parties, one under the immediate supervision of Stevens, to start from St. Paul's, and the other under General, then Captain McClellan, whose orders directed him to repair *via* the Isthmus of Panama to the Pacific, explore thoroughly the passes of the Cascade Mountains and the pass of the main Columbia for a railroad line. The field to be explored was new, and mostly untouched. It is true Long and Nicolet had explored the region to the sources of the Mississippi;

3*

but here we expected no physical obstacle in the location of our line. It was only when reaching the Rocky Mountain region that we anticipated meeting with nature in her sterner and more forbidding features.

Our parties, after covering as broad a belt of country as practicable with the means at our disposal, did not arrive at the Rocky Mountains until late in the autumn of 1853. But our mission here was far from being completed. The mountains were expected to prove the key of our troubles; to leave them unexplored, left us unarmed in knowledge as to the real merits of the difficulties that must needs be discussed in the great question of a railroad line to the Pacific. It was therefore determined by Stevens to leave a small party in the Rocky Range to continue the exploration of the mountain system, to map the passes and approaches of the mountains on their either slope, to trace the headwaters of the Missouri and Columbia, to ascertain the depths of the snows, and make known such facts, geographical, meteorological, and statistical, as should come within the field of our observation. To the charge of this party I was assigned in the autumn of 1853, and continued on duty with it until February, 1855. The mountains, for eight degrees of latitude and nine degrees of longitude, covering an area of 232,-000 square miles, were traversed in many directions. The branches of the north and south forks of the Columbia, with their tributaries were traced, and the exact position of their headwaters determined. The same was done with reference to the headwaters of the Missouri, and for the first time in the history of the geography of the country did we have a complete and authentic outline map of the Rocky Mountain regions, where rise the headwaters of two great arteries of the continent, the Missouri and Columbia. I established my headquarters in a mild, sheltered valley of the mountains, where erecting rude log huts for the comfort of my men, I made this the centre of geographical explorations.

During one of these tours in the spring of 1854, I traced one of the branches of the Clark's Fork to its source in the Rocky Mountains, where I discovered a pass so low and practicable that I reported the facts to the War Department, stating that with a moderate amount of labor, a first-class stage-road could be here constructed, and gave the *experimentum crucis* by taking a wagon train through it on my return across the mountains in March, 1854.

In this latitude the main chain of the Rocky Mountains is much less difficult than some of its spurs. This is particularly so with the Bitter Root Range, which, though it does not attain the same altitude, is in winter covered with a depth of ten feet of snow, and offers so many obstructions to continuous travel, that it has become the dread and terror of those whose duties compel them to cross this range in winter. We can avoid this depth of snow, however, by going farther to the north. Paradoxical as it may appear to some, it is nevertheless true, that within reasonable limits in latitude 47 and 49 degrees we find this phenomena to hold, that the farther we go north the more mild the climate becomes, less severe the cold, less deep the snow. This is shown particularly *via* the Clark's Fork of the Columbia, where one can travel during the most inclement winter, whereas by the pass sixty miles to its south, we find the mountains blocked by snow as early as December, and continuing until the 1st of April. The main chain of the Rocky Mountains here has an elevation of from 6,000 to 7,000 feet above the level of the sea, yet I have never found the snow deep enough upon them to render travel impracticable during the winter season. The Indians who inhabit these mountains, cross and recross the range in quest of buffalo every month in winter, and though they may be delayed by a severe snow drift, or an occasional deep snow, it is but seldom that they allow themselves to be snow-bound. Their horses are of a peculiar hardy nature that withstands the phases of winter, subsists solely upon the native grasses found along the hillsides in winter, and in the valley bottoms in summer. I have often compared the perfectly careless abandon of these people, so far as the care of their stock is concerned—for they have horned stock as well as horses—with the sedulous care that the farmers of New England and in the Northwestern States evince in procuring and husbanding forage for their stock, during a dreary six months of winter. It is this important fact of mild and genial winters, with rich and nutritious grasses, that renders this region one of the finest grazing sections to be found in the world.

I have never seen in any region finer or richer flavored beef. Even in the most severe winters, I have seen our beeves so fat in these Rocky Mountain valleys that we could scarce eat the meat.

During my explorations in the winter of 1853, my attention was specially called to one fact—a meteorological phenomenon as useful now as it was then, which is destined to have a most marked bearing in the future settlement and development of the eastern slopes of the Rocky Mountains. It is what I have termed an atmospheric river of heat. I first noticed this feature in the Beaver Head Valley, on the Jefferson Fork, a region which to-day is being rapidly peopled by a hardy, thrifty class of miners, and which point in time is destined to be a marked geographical and populous centre. This river of heat seems to begin somewhere south-

west of the Black Hills in Nebraska, and coursing along the
eastern base of the Wind River Mountains, crossing the
tributaries of the Upper Yellowstone and Missouri Rivers,
reaching the main chain of the Rocky Mountains in latitude
46° and longitude 113°, then crossing the range follows the
Hellsgate River to the Clark's Fork of the Columbia, and
thence along the valley of this tributary, till it diffuses itself
in a fan-like shape on the plains of the Columbia. Its width
is from one mile to two hundred miles, depending upon the
configuration and character of the face of the country over
which it passes.

It warms up the entire eastern slope of the mountains,
giving this region pleasant and genial winters. I have more
than once asked: Why the name of the Wind River Mountains in this section of the general Rocky Range ? Has it
arisen from the presence of this marked wind or Wind River
of heat, or from some other cause ? The Wind River Range
rising from the Laramie Plains in latitude 42½° and longitude 109°, where it has an elevation of 6,000 feet above the
level of the sea, trends northwestwardly as a curvilinear
wall, with its concavity turned towards the northeast, gaining in altitude, till it has attained at Fremont's Peak an elevation of 13,000 feet. Its altitude falls off gradually, till in
latitude 45°, it has an elevation of 8,000 feet above the level
of the sea.

This atmospheric river of heat courses along this range,
and in its passage traverses a region where are located a
number of hot sulphur springs, and where exists on the Big
Horn River a coal-oil spring, similar in all respects to the
coal-oil springs of Western Pennsylvania. This phenomenon, I say, is as curious as it is useful, in pointing out a new
route of travel, and passing as it does over a region capable

of settlement, already locates upon our chart of the settlements of the Rocky Mountains, a line of population from
Fort Laramie to the headwaters of the Columbia River.
When it is remembered what marvellous results and conclusions have been arrived at, and what deductions made in the
study of the isothermal laws, which are here especially and
distinctly marked, we shall not be surprised, at no distant
date, to be enabled to trace a line of settlement from the
valley of the Platte to the plains of the Columbia, with a
people holding kindred views, filling kindred occupations,
and aiming at a kindred destiny.   My mission of exploration in the central section of the mountains being completed
by the autumn of 1854, I returned to the Pacific, tracing on
my passage the route followed by Lewis and Clark in 1804,
and must say that I found it to be the most difficult pass
that it fell to my lot to examine in the whole mountain system.   The entire field explored by us during the period of
two years, developed the feasibility of the country for a railroad line, from the great chain of Northern Lakes to the Pacific, and *via* the valley of the Columbia.   Though we did
not succeed in securing Congressional action in behalf of a
northern line, we did succeed in attracting sufficient attention
to the region to inaugurate the project of opening up wagon
and stage communication from the Missouri to the Columbia
Rivers.

On returning to the City of Washington in 1855, a full
report of my labors was laid before the War Department,
which set forth that a new route of travel could be had both
for military and emigrant operations over a line direct from
the Columbia to the Missouri, in a distance of six hundred
miles—by the expenditure of a moderate amount of means.
The project was laid before Jefferson Davis, then Secretary

of War, but the want of adequate means at that date did not allow the work to be favorably acted upon, and hence nothing was done until the winter of 1857, when I was again ordered to Washington, and the Government resolved to put into execution the project of a northern overland communication, directing me to superintend the construction of the same.

This was commenced in the spring of 1858, and completed in the autumn of 1863, a period of five years. In its prosecution we were compelled to pass three winters on the road. Our line from the main Columbia sought for its location the valley of the tributaries of that stream, as far as their position lent themselves to our proposed direction, till reaching the summits of the Rocky Mountains, traversing which we crossed the tributaries of the Missouri, following along its northern bank until we reached the head of steam navigation on that stream at Fort Benton. The country, through which we traveled for two hundred miles, is a high rolling prairie, or table land, 2,000 feet above the level of the sea. When we reach the western foot slopes of the Bitter Root Range, or as it is termed by some the Cœur-d'Aléne Mountains, we find ourselves in this bed of mountains for 300 miles, following lines of water-courses and making practicable locations wherever natural ones did not offer. Crossing the main chain of the Rocky Mountains in latitude 47°, we take advantage of the minor and lower spurs of the main range, and at once enter upon the broad swelling prairies or plains of the Missouri. The line of location as traveled by us is as follows:

On leaving the Columbia at Wallula, or Old Fort Walla-Walla, we pass up the valley of Walla-Walla to the New Fort; thence to Dry Creek; thence to the Touchet; thence to Snake River, at the mouth of the Palouse; thence along its valley to the Cœur-d'Aléne Lake; thence to the Cœur-d'Aléne Mission ; thence to the St. Regis Borgia, and by its valley to the Bitter Root; by its valley to the Hellsgate; by its valley to the Little Blackfoot; by its valley to Mullan's Pass; thence to the Big Prickly Pear; thence to Fir Creek ; thence to Soft Bed, Silver and Willow Creek, and the Little Prickly Pear, to the Dearborn River; thence *via* Bird Tail Rock to Sun River ; thence *via* the Lake to Fort Benton.

I found, however, that a section of thirty miles was so wet in early spring, due partially to the overflow of the St. Joseph's River and the junction of many small lakes, that I was forced, during the year 1861, to change the location, at the expense of thirty miles of new road.

I will not dwell upon the special details of this work, which cost much labor and anxiety, not only in the actual work performed, but in the varied questions of direction, location and judiciousness of labor to be expended. But I would simply state, that in this line of 624 miles of road we cut through 120 miles of most dense forest a width of thirty feet; 150 miles through open pines, and 30 miles of excavation in earth and rock, occupying a period of five years, and at a cost of $230,000.   In addition to performing this special labor, we explored and surveyed the country to our either side, covering a belt or zone of three hundred miles.   By this means we were enabled to trace the headwaters of the Columbia and Missouri Rivers in such a manner, that though we left much undone, we yet collected sufficient material from which to complete a very correct outline of the Rocky Mountain system, with its vast network of spurs, streams and tributaries.   The main forks of the

Columbia have their rise, one in British territory, in latitude 52° north, and the other in the Wind River Mountains, near 42° of latitude, thus draining a line of ten degrees of latitude, and reaching the coast in fourteen degrees of longitude.

The Missouri has its rise in longitude 112°, and in a line of 17° of latitude. In a distance of seven hundred miles this stream flows nearly due east, along the forty-eighth degree of north latitude; when reaching the 102d meridian it trends south-westwardly, and continues on this course till it has reached its reservoir, the Mississippi. It will, therefore, be seen at a glance that the Rocky Mountains constitute the great heart of our interior, from which arteries and veins flow for thousands of miles, pulsating with life, activity and vigor, and upon the bosoms of which are floated the golden products of the bowels of the Rocky Range on the one side, and the rich offerings of Ceres on the other.

Any one who will study the geography of the North Pacific will observe this fact, that the Columbia is the only stream which, rising in the main chain of the Rocky Mountains, breaks through the coast range that extends from British Columbia to Mexico, and empties into the ocean; that its headwaters and tributaries interlock with the headwaters and tributaries of the Missouri; so that, standing on the summit of the mountains at any point, you can see the waters that flow into two oceans, and that nowhere on the continent do we find such a perfect network and ramification of water-courses as we find here. The large volume of water, giving off heat, as it does, must necessarily tend to temper the climate of the mountain valleys. This we find to be the case, as evinced in the uniformly mild and genial winter climate, where stock can graze on the side-hills in winter without any forage being provided for them. An-

other feature that the geographer and explorer develops in
this region is the infinite number of sheltered valleys found
embosomed in the mountains. These valleys constitute the
homes and abiding places of the Indians, and promise to be
important nuclei in the settlement of the country. These
valleys are all more or less connected together by natural
wagon roads, and are already indeed being taken up by farmers and graziers.

The Missouri River is navigable for 3,100 miles from its
mouth. When we reach the great falls of that stream, there
for thirteen miles the river, in a series of cascades, falls,
chutes and rapids, has a total fall of 380 feet. The land to
the north, and for four or five miles back from the river, is
much broken by coulees and ravines, but to the south, and
distant three miles, the country is a flat plain, affording
every advantage for a first-class stage line, and over which
a portage of seventeen miles can be had, when we reach the
head of the falls, where the river, in a broad placid sheet,
looks like a silvery lake. The banks here become low,
fringed with cottonwood, and from this point we can have
175 miles of further navigation, which will bring steamers
near 3,000 miles from St. Louis, and to the Beaver Head
Valley in Montana Territory.

The Yellowstone, one of the principal feeders or tributaries of the Missouri, is said to be navigable for nearly 700
miles. Capt. Maynadier, an energetic and capable officer,
made a survey of this stream three years ago, and both he
and General Warren, an old and efficient explorer in that
region, are of the opinion that this large river is eminently
practicable for steamers. Should this be the case, 700 additional miles will be added to the 3,000 mile stretch of the
Missouri. There are a few tributaries to the main Missouri

where steamers might be used to advantage for short distances, and among others I would mention the Sun or Medicine River, the Jefferson Fork and probably the Gallatin and Madison Forks of the Missouri.

The country bordering these last mentioned streams is among the most beautiful to be found west of the Mississippi. The country is a gently undulating prairie, dotted here and there with clumps of timber. All the streams are beautifully fringed with forest growth, the soil is rich, climate mild and invigorating, and all the elements for happy homes are here to be found. Indeed, I do not speak my own experience alone, but that of all whose fortune it has been to traverse this great area, dating from the exploration of Lewis and Clark to the present period, and this experience is that no region in the far west offers such inducements to the farmer or grazier, or sites for more beautiful, pleasant, or healthful homes, than at the three forks of the Missouri River. When the entire country was a wilderness, and occupied solely by Indian bands, it was both folly and madness to suppose that permanent settlers would wend thither their way with a view of making homes, and especially when there was no market for the produce of the soil. But when the hardy miner had prospected the hidden golden resources of the bowels of the mountains, and made known the wealth of this new El Dorado in our midst, inducements of no ordinary character were offered the farmer and grazier. Numbers now constituted security, and the mines a market, until this great system of mountain valleys are to-day teeming with an industrious, energetic, and laborious population, and the newly-organized territories of Idaho and Montana promise at no distant date to be incorporated as the Golden States of the Confederacy. The Columbia River is navigable with two portages, both made now by rail, for 450 miles from its mouth, where we reach its South Fork, which steamers have navigated as far as Lewiston, the present capital of Idaho, running from early spring to late winter, and the proposition is now on foot to add three hundred miles more to its navigation by running steamers as far as Fort Boise, for the double purpose of supplying a military post, which it is proposed to build there, as well as the mines recently discovered in its vicinity.

Many are sanguine that the project is feasible, and for one I have no doubt but that the experiment will be crowned with success. These streams are seldom blocked by ice, only during the severest winters, and then only for a short period, so that we might say we have a season of uninterrupted navigation. Many towns and villages have already sprung up along the rivers, the principal one of which is Lewiston, named in honor of Captain Lewis, the pioneer explorer of 1804. It is rapidly increasing in wealth and importance, has a vigorous, industrious, and capable class of citizens, who have taken the lead in all those developments that look towards the early and rapid settlement of the country, and being the capital of the newly-organized territory of Idaho, must at an early date be a city of note. In the interior we have the cities of Florence, Elk, Oro, Fino, Pierce, Millersburg, Auburn, Placerville, Idaho, Centreville, Hogam, Boisé, Esmeralda, Silver City and Owyhee, all nuclei in rich and extensive mining districts. On what we term the Upper Columbia we now have twelve steamers running, all of which debark their freight at a landing on the Columbia, called Wallula, or Old Fort Walla-Walla, from which point everything is freighted by wagons and pack trains into the very heart of the mountains. From a survey of the main Columbia, we are led to believe that with few portages this stream is navigable for many miles northward, probably to and beyond the 49th parallel of latitude. But as yet the mineral wealth in this quarter is so limited, so far as discovered, that not sufficient inducements exist for private enterprise to test the river by steam. The Oregon Steam Navigation Company is the pioneer steamboat company in this region, and deserves much credit for the bold and vigorous prosecution of all those projects that have so greatly added to the development and the wealth of the interior of the Pacific Slope. As soon as inducements shall offer, this company will have their steamers ploughing the waters of the main stem of the Columbia, even into British territory. The agricultural resources of many sections of the North Pacific, and especially in the interior, compare favorably with those of California, where everything would seem to grow under a climate so varied, that we find the grape and the orange blooming in sight of perpetual snow and ice. The Cascade Range of Mountains, which in Oregon and Washington are an extension of the Sierra Nevada and Sierra Madre Mountains of California, run parallel to and distant from the ocean line about 300 miles, and rising to an altitude of 5,000 to 6,000 feet above the level of the sea. This ridge or backbone of mountain system acts as an effective wall or barrier in the rainy season to throw back the moisture and rains, so as to give to the interior a totally different climate.

As is well known, we have all along the coast from the Tropics to the North Pacific a wet and dry season, continuing each for six months. But this line of mountains rising to a higher altitude than the rain clouds, prevents the moisture from passing over to the interior plains of the Columbia, and the consequence is we have an equable and uniformly pleasant temperature during the entire year.

During the past spring the farmers at Walla-Walla were ploughing on the 8th of February, and planting on the 15th of the same month—this in latitude 46° north, while we in latitude 38° and 40° were still blockaded by a long, dreary, and threatening winter, from which we have now only emerged. Wheat, oats, barley, and corn are here produced in rich abundance. Wheat 30 to 40 bushels to the acre, barley and oats the same, and corn 80 bushels to the acre. Apples, pears and peaches, here grow well, and both climate and soil would seem to favor the largest yields. Potatoes yield 400 to 600 bushels to the acre, and I have seen 800 grown. The market for all this produce at present is in the mines, where the demand is so great that California is called upon to supply that which the farmers do not produce.

The section of Oregon and Washington to the west of the coast range, with the exception of the meadow or prairie bottoms, is a dense forest, timbered with the fir, pine, and oak — the last being of a scattered and mostly scrubby growth, and not suited for or used in the mechanic arts, and hence one of the great commercial interests along the Columbia and Puget Sound is that of the timber trade. The lumber is shipped to China, Japan, Australia, and the Sandwich Islands. The finest spar and mast timber to be found in the world is here grown, and shipped to all quarters of the globe. It is from here that the shipyards of France and England are largely supplied with their spar timber, and to all quarters of the globe is a similar trade being rapidly built up. This, together with the great coal trade from Bellingham Bay, and the fishing interests of Puget Sound, is destined to render this great inland sea one of our richest possessions in the North Pacific. The agricultural scope of country bordering its waters is not large enough upon which to base large expectations; but this interest, small though it be, when compared with others, must tend to swell the geographical importance of this great commercial door, through which eventually will pass the trade from the East and the isles of the Pacific to our Atlantic seaboard.

The country to the interior or between the Coast and Rocky Range is more flat or prairie, and though it is not continuously cultivable, we find many agricultural areas, where every want of man can be supplied. There has existed a fallacy as to the true agricultural capacity of this quarter which exploration and examination have radically dissipated. This applies not only to the section west of the Rocky Mountains, but to a large area to the east of this range, and extending on both sides of our northern boundary.

In reading the reports of Captains Palisser and Blackstone of the English Army, and that of Dr. Hector, the English Geologist, I noticed a remarkable feature to which they refer with special stress, and which they have delineated on the maps that accompany the reports of their explorations through the Sascatchewan country and the region lying between Lake Winnipeg and the Rocky Mountains.

Extending to the south of our boundary, they discover a broad belt of fertile soil of the first quality, where the winters are represented as mild, and where the Hudson Bay Company have produced every crop grown in our Northwestern States. In one body Captain Pelisser, an old and experienced traveler, estimates 11,000,000 acres of beautiful soil. I was astonished at the facts and figures as given in his report published by the House of Parliament three years ago. In conversing on the subject with Dr. Whittlesey, of Cincinnati, who has spent many years in the geological examination of Northwestern States, he confirmed the

views entertained by these English travelers, and in giving
me his conclusions of the agricultural capacity of this scope
of country westward from Lake Superior to the Rocky
Mountains, he says, in a letter to me of the 23d March,
1863 :—" The following is a succinct statement of my views
in reference to the portions of the Western States to which
we must look for a permanent supply of wheat. In the fall
of 1848 I was at Red Lake in Northern Minnesota, when
Mr. Ayres, a very intelligent missionary, came in from the
British Settlements on Red River. He brought some un-
bolted flour from wheat grown upon the banks of that river,
at the Pembina Settlement. It was the sweetest and most
nutritious flour I ever ate. He stated that the employees
of the Hudson Bay Company had for a quarter of a century
produced all the wheat they needed, that it was a sure crop,
the grain sound, and that it would yield 40 bushels to the
acre. The result of my examinations at that time on our
northern frontiers, was a conviction that the true wheat-
growing country of the United States lies north of the St.
Peter's River, and west of the Mississippi. Since then the
State of Minnesota has settled rapidly, and in confirmation
of my views the Commissioners of Statistics for the State
reported, in 1860, that there was then a surplus of (3,000,-
000) three millions of bushels of wheat."

The agricultural statistics of Ohio, Indiana, Illinois, Mich-
igan and Wisconsin show that wherever corn is a thriving
crop, wheat is a diminishing one.

In Europe, wheat is more profitable and sure in those
countries where it is too cold for Indian corn. On this continent, I am convinced that the territory north of our boundary, including the valley of the Sascatchewan River and
Lake Winnipeg, is destined to be a region of wheat. I have

sometimes thought, if Great Britain covets the Cotton States, we might make a good bargain by exchanging them for this great field for cereals, to which our people would emigrate in great numbers if it was ours.

It will not be long in the future when wheat will be as much more powerful than cotton as food is more necessary than raiment.   Though I do not fully coincide with the Doctor in reference to his rate of exchange or the feasibility of the barter, yet I do think that in the settlement of our north boundary question we actually gave away what the Doctor would now barter back, and that in this adjustment we were in a great measure shorn of our territorial heritage.

In case of a war with our neutral friends, we may yet be enabled to send them breadstuffs raised from a soil where the English flag now floats in triumph.

With a Russian civilization on one side and an American civilization on the other, the day cannot be far distant when this territorial sandwich will be devoured by the one or the other, or jointly, at the great feast-table of national necessity.

The map of the region would also show another feature, which, in the gradual settlement of the country both by Americans and Englishmen, must play an important part in the commercial economy of the interior of the Northwest Pacific.   This is the number of large and navigable rivers, which, rising in the Rocky Mountains, pour their tribute eastward, near our own boundary, and finally find their reservoir in Hudson's Bay.   The principal of these is the Saskatchewan, with its two main forks rising in the main range, and flowing with navigable stretches for near 2,000 miles. These rivers, with a proper system of connection by locks and canals with the chain of great lakes to their east, would give us water communication from the St. Lawrence to the very

4

bases of the Rocky Mountains, and is a project worthy the
earliest and earnest attention of those who would connect
New York with China and Japan by a route across our own
continent.

The field explored by Palisser and Blackstone connects
with our own, and hence, in my general map, accompanying
this, I have extended my limits sufficiently far northward
to include, partially at least, their new and interesting field
of labors.

The principal object of the exploration of these gentlemen,
which was inaugurated in a measure under the auspices of
the Royal Geographical Society, was to discover, if practicable, a route for a railroad line to the Pacific, in order to
connect West British Columbia with the Canadas. Their
labors, though enlarging greatly the field of Geographical
Science, failed in developing the feasibility of such a route :
the physical difficulties of the country drove them nearer
our boundary than they expected to travel, and finally compelled them to cross it, in order to reach the Pacific, thus
proving conclusively that no pass exists in the Rocky Mountains where the iron rail will be laid north of the 49° : if
there was a pass in the main range, the spurs and mountains
to the west of this range are so broken and difficult as to preclude the possibility of a direct route to the ocean. Those
who will study the geography at the headwaters of the Columbia River will not fail to observe this marked feature,
that for many miles from its sources in the mountains, the
principal fork of the Columbia, rising in British Columbia,
flows along the bases of the mountains first northwardly, till
finding a practicable gateway through the mountain defiles, it
flows southwardly as an immense canal, and in echelon, and
it is only after crossing our northern boundary that the phy-

sical character of the country enables it to tread westwardly towards the Pacific, which it does through the natural gateway of the Cascade Portage, the keys of which, fortunately, are in our own possession.

Outside of their agricultural capacities, our North Pacific possessions attracted but very little attention until 1858 ; at this time much to the surprise of even experienced miners, gold discoveries were made on Frazier's River, in British Columbia, that attracted to them large numbers of miners from California and Oregon. The country was but little known ; there was little disposition to explore, outside of the immediate gold localities ; in addition to all other troubles, the charter of the Hudson's Bay Company expired just at the time, and the project under discussion before the English Parliament, was either to renew the charter of this company, or to give a colonial organization to British Columbia. While the matter was involved in discussion, and all action was kept in abeyance, the disposition of the Hudson's Bay Company was to deter miners, and especially Americans, from either prospecting or entering the country.

The friends of the Colonial project succeeded ; British Columbia was organized, and the country thrown open to settlement both to Englishmen and Americans. The door of settlement once opened, the hardy miners from California and Oregon ramified in every direction, until the mines were thoroughly prospected and worked ; until steamers ploughed every navigable stretch of water ; until wagon roads and stage lines were in successful operation ; until villages, towns, and cities reared their heads in the wilderness ; until Victoria became a free port and a rival of San Francisco, and Vancouver's Island was about to become a golden possession to the English Government ; and British Columbia bids fair to-day to become our peaceful friend and ally, or our dangerous enemy and neighboring foe.

It was while the subject of routes of travel to and from these mines was being discussed, that a shorter and more economical route of travel to them was found—it was while the miners where passing to and fro across the British border in 1858, that gold discoveries were first made in Washington Territory, on the eastern slope of the Cascade Mountains, and on the minor tributaries of the Columbia. The miners worked these localities until hostile bands of Indians drove them off, and until they were compelled to search in quest of new discoveries. Gold continued to be found in 1859 and 1860 along the Upper Columbia and its tributaries, and in the summer of 1860 the first gold discoveries were made in the Nez Percés mines, on the western slopes of the Bitter Root Mountains, by a party under Captain Pierce. These mines were worked until 1861, when the miners in their prospecting tours discovered the Salmon River mines, and continued their search over the mountains until they have successively discovered and are now working gold mines on the John Day's Powder River, Grand Rondo River, Burnt River, Boisé River, Payette River, Salmon and Clearwater Rivers and Owyhee, all on the western slope of the Bitter Root Mountains. In and on the east slope of the Bitter Root Mountains my party had found gold as early as 1853, but not in sufficient quantities to warrant parties working the mines. Discoveries were again made in 1860, and in the summer and autumn of 1861 for the first time did the miners feel authorized to go extensively into mining operations. It was not long, however, before we found parties out prospecting in all directions, and now mines are being successively worked in the Deer Lodge Valley, Beaver

Head Valley, Big Hole Valley, and the Prickly Pear Valley, at the headwaters of the Columbia and Missouri Rivers. It would be safe to say that we have a population of 50,000 men now engaged in developing the wealth of these mines, the annual yield from which may be safely estimated at $20,000,000. As yet no traces of silver have been discovered, but when we compare the features of the great interior basin of the Columbia with those of the Colorado, we are so forcibly struck with the many elements of similarity between the two that we make bold in saying that what we have discovered in the one, we have a right to anticipate discovering in the other.

Though it has pleased many persons, for reasons which I am not charitable enough to think were even satisfactory to themselves, to term the great plain of the Columbia River an immense desert, I am still sanguine to believe that in this same plain or so called desert we shall find as rich a wealth as the desert of Colorado is now sending forth to the commercial world. This desert, and the river flowing through it, for its agricultural capacities alone has been favorably compared by Judge Edmonds, of the General Land Office, to the Nile, the enriching influence of which has made Egypt the granary of the East since the earliest period of man. In its primeval days, twenty millions of people dwelt in that region, and tilled its spacious and fatty acres. These people did not limit themselves to what nature had done for them, but by a mighty system of irrigation, which even invaded the Lybian desert, rescued from the dominion of burning sands extensive tracts of country, making them arable by throwing upon them enriching waters, which the industry of man effected, through canals, lakes and accequaics, some a hundred miles in length, and counted by hundreds and thousands. Their fertilizing power, as stated by Edmonds, is shown in the fact that under the Pharaohs, the rule of the Ptolemies, Roman or Mohammedan, Egypt has paid even as a colonial dependency immense tributes to the support of its people and their arbitrary rulers. The memorials of its triumphs still exist. Its piles of architecture, covering acres and rising to the clouds—its avenues of sphynxes, colossal figures and archways—are there. We have a right, therefore, to take a lesson from the past, and for one I am fain to believe that what has proved true in regard to Egypt and the Nile, will prove equally true of California and Colorado—of Washington Territory and the Columbia—and that we shall yet see a bold, numerous, thrifty, wealthy population peopling the vast area lying between the Rocky Mountains and the Pacific coast line. In our exploration through the mountains we discover other minerals besides gold. We find the red hematite iron ore, traces of copper and plumbago. Then along the main Bitter Root River we also find iron ore. Also along the Clark's Fork. Lead for miles is found along the Kootenay River. Coal is found at the Three Butts, sixty miles northwest of Fort Benton, and it is also said to be found on the Clearwater River. It is also traced throughout the length of the Missouri River. We find limestone in great abundance in the Rocky Mountains, on the headwaters of the Columbia, and on the Little Blackfoot. Also on the Hellgate, the Clearwater and Touchet Rivers iron pyrites were found at many localities, and I am sanguine to believe that when the mountains are thoroughly explored we shall find the mineral wealth of this region to compare favorably with that discovered in California and Nevada. This singular feature would seem to hold both n the Rocky, Sierra Nevada, and Cascade ranges, that on both slopes of the mountains we find similar mineral deposits at the same altitudes above the level of the sea. The Indians on the headwaters of the Missouri have reported the existence of cinnabar, and from the reports made by General Lander ten years ago, I am sanguine to believe that we may look for rich quicksilver mines on both slopes of the Rocky Mountains, in the regions where the gold mines are being now opened.

INDIANS.

Within the area which it fell to our lot to examine, it may be soberly stated that from 30,000 to 40,000 Indians are found. Their habits are of the same nomadic character that mark other portions of the race on the continent. They are mostly friendly, except in Eastern Oregon, where the Shoshonee and Bannocks reside. Though the presence of troops has done much to give security and confidence both to frontier settlers and emigrants, yet we are not perfectly free from the occurrence of those massacres which have caused humanity to shudder.

The policy of the Government with reference to the Indians has undergone but little change in our relations with them since the earliest settlement of the country by the Pilgrims, who found a continent at their feet, rich in lakes, rivers, and in all that was requisite for the happy habitation of man, as truthfully and beautifully told by " Observer," writing to the *Oregonian :* " Savage tribes roamed more or less over this fair heritage solely for fishing and hunting. The feebleness of the colonists necessitated treaties and purchases from the Indian tribes, and a policy, originating in our weakness, has been maintained and continued in our strength. The Government recognized the right to the soil in the wild, untutored savage, and treated with him as a political equal; and the philanthropist and Christian looked upon him as a ward committed to their care to be civilized and christianized. It was a beautiful faith—a grand dream —soon to be dissipated, however, in the practical workings of our civil and political machinery. It was a governmental error in thus conceding to savage hands such rights to the soil. The earth was made to be tilled, and made fruitful even to the maximum degree; and if, in its subjugation, settlement and cultivation, the fish should disappear from its creeks and rivers, and game from its forests, they were incidents to civilization for which no savage tribes could claim compensation. The Government, in its liberality, has proposed to teach the Indian the arts of husbandry, and to furnish him with all the implements and appliances of the husbandman; but the scythe hangs unused at the agency—the axe lies still unhelved—and the ploughshare rusts at his cottage door. The Indian still dries his salmon on the banks of the silvery streams that glide by his lodge—still digs his roots from the prairie where nature planted ages ago—and still resorts to the buffalo chase in quest of the bison that roam as yet in millions over the western plains—and when his toils are ended and his wants are supplied, in his mat-constructed hut he throws himself down to rest upon his furs and skins. The school-house opens in vain its doors to him, for he despises her letters. In the varied book that nature spreads out before him he learns his lessons; and his poetry, if poetry he has, he drinks from the heavens when sentinel stars keep their watch in the night. The missionary has gone to him with a heart overflowing with kindness and

Christian love; but whatever balm the Bible may possess,
it has borne on its wings no healings to the hut of the Indian. With an apathetic, confused, indefinite and dreamy
faith he looks for fairer hunting-grounds in the spirit-land,
where the streams abound in salmon, the woods are filled
with game, and where his every material want is supplied
by the hand of the Great Spirit who directs him thither.
The experience of those who have seen most of the Indian
has been neither flattering to the efforts of the Government,
nor consoling to the hopes of the true Christian philanthropist; but the purposes of the red man's creation in the economy of nature are well nigh accomplished, and no human
hand can avert his early extermination from the face of the
North American continent. Silently but irresistibly the
purposes of Providence take their way through ages, and
across the line of their march treaties would seem but
shreds, and the plans of man on the tide of history but waifs
upon the sea."

The first printing press introduced on the Pacific, within
the limits of our possessions, was brought thither in 1836,
by a missionary of the name of Spaulding, and put in operation among the Nez Percés Indians, in the new territory of
Idaho. A log hut still stands to mark the place and the
spot dedicated to the diffusion of knowledge. Flowers
bloom about the cabin, trees of his own planting here yield
their delicious fruits. Since that date how the press has
advanced in power on the Pacific Coast! Throughout its
length and breadth sheets of light now gleam, and its voice,
weak and feeble then, but mighty and powerful now, is
heard not only in the cabin of the pioneer, where in the as
yet unbroken forest the smoke curls up to Heaven, but in the

Cabinet Council of the Nation the Chief Magistrate is forced
to catch its notes from that distant shore. But change has
not only stamped the press, but has invaded the lethargy of
all nature in her deep sleep, even down in the very bowels
of the hoary-headed mountain. The waters, on the bosom
of which the canoe of the savage yesterday quietly glided,
are to-day vexed by the prows of heavily freighted steamers, as they plough and furrow their depths of blue; the
farm blossoms in the valley, the rude hut erects its head to
dot the entire land, and the woodman's axe is heard in the
primeval forest; villages, towns, and cities have sprung up,
as if by magic, and the visible fruits of the invisible alchemist surround us on every side. From every harbor and
roadstead, from every inlet and indenture along the coast, the
white sails of commerce are spreading their pinions to the
breeze, and where yesterday stood bare, bleak, brown sand-
hills, with the rude village of Yerba Buena as their pride,
to-day rises a cupolaed city, with a hundred thousand beat-
ing hearts to give new life to a region redolent with the
songs of thrift and plenty, rising from the waters of a majestic bay, at the threshold of a golden gate, wearing a cor-
onet upon her brow, adorned with jewels gathered from the
beds of her own rivers and creeks, and glittering with gems
plucked from the bowels of her own noble Sierras. Glorious, Golden State, vigorous and active, healthful and generous, young and ambitious, with one hand laying the iron
track on which to send you her trade and her treasure, and
with the other stretching her commerce across the seas, to
grasp the glittering and opulent wealth of the Indies and the
East. May thy future be as great as thy past has been
poetical, and may thy sons and daughters be proud to own
thee for so noble a mother!

But we are not dependent exclusively upon the press for our intelligence from the East; for the telegraph having placed New York and Oregon under the same meridian in point of time, we find the thought, as uttered to-day in Wall street, printed in the *Pacific Bulletin* by to-morrow's sun. The telegraph to us is now a material necessity, and already are we extending our lines northward, connecting with Puget Sound, British Columbia and the mining localities in the eastern section of Washington and Oregon, thus preparing the way for the Russian connection, which must some day be *un fait accompli*. St. Petersburgh and San Francisco will then hold their *matinées*, and their friends at the Aleutian islands afford a resting spot for the conferences of Russia and America. This once accomplished, San Francisco will contend with New York for the mastery of the intelligence of the world; for already, by political anatomists, she is pronounced the brain or cerebral centre of the earth, where all the terrestrial nerves report to be re-distributed. London and Paris will be but ten days distant. Hence, if it is not so now, it must soon become to us a fixed necessity. It can no more be permanently destroyed, when once completed in a circuit around the globe, than the grown oak can be crowded back into the acorn. The long anticipated luxury, once enjoyed, becomes a necessity that must needs be gratified.

ROCKY MOUNTAIN MISSIONS.

Among the marked features found in the wilderness region of the Rocky Mountains, one will not fail to note that of the presence of the Catholic missions, established for the benefit of the Indian. These establishments, so many St. Bernards in the dreary mountains, dispense their kindness and hospitalities not only among the Indians, in whose midst the Cross has been thus erected, but to the weary traveler and emigrant, as he wends his way to the shores of the distant Western ocean. These noble fathers have resided in the country for nearly half a century, endeavoring to reclaim the Indian from the wild nomadic life which seems to be a part of his very being. Though the efforts of these zealous Jesuits have been great and untiring, I regret that the results of their labors have been as few as they have. The Indian has but seldom appreciated the poetry that surrounds the erection of the Cross in the wilderness for the salvation of man. His life is devoted to the material things of this world, and drinks in no poetry from the teachings of the Gospel. When the buffalo shall cease to give him sustenance, the Indian will disappear from the face of the continent, and live only in the records of the past.

Though the field of our exploration was mostly confined to the Rocky Mountain system proper, still there was a large area east of the mountains which it fell to our lot to explore, and though here and there large tracts of good soil were to be found fit for cultivation, yet the major part of the soil lying between the Missouri and the North and the Big Horn River is an immense bed of *mauvaises terres* unfit for cultivation, and upon which for miles nature does not smile with either a tree or a shrub, and where not even a blade of grass seems disposed to grow. This tract begins to the North of the Niobrara in Nebraska, and continues to the Yellowstone, and westward to the more eastern spurs of the Wind River Mts., and is now and ever will be, I think, unsuited to the habitation of man. Though this region acts as a break to the continuous settlement and population of the country, yet it does not act as a barrier to a direct communication to the Pacific from the Northwestern States.

In the study of this region, I have been forced to the conclusion that our best Northern route across the continent would be to start out from Lake Superior and take a line direct from Fort Clark on the Missouri, and thence to the Valley of the Yellowstone, and to follow this valley to its sources in the Rocky Mountains, and thence cross the range either by the Valley of the Jefferson Fork and Deer Lodge, or *via* Salmon River and the South Fork of the Columbia. A very special examination, however, should be made *via* this route, and its feasibility practically determined. The geography of the country would seem to point this line out as feasible, and passing as it does through the mineral region of the central sections of the Rocky Mountains, would become one of much agricultural and mineral development. When the question is to be determined what route it is best to follow in order to reach the North Pacific by means of the iron rail, this region will occupy a large share of the attention of the railroad world. For one, I have always contended that the Valley of the Columbia should not be ignored in any line of location that should seek the Pacific; its position and the great mineral resources now found along its headwaters, all warrant us in thinking that either this valley will enjoy an independent northern route, or be the location for a branch to the present central line from St. Louis to Sacramento, and when the beautiful broad entrance of Puget Sound, with its magnificent proportions of an inland sea, are compared with the difficult entrance of the Columbia, where lines of breakers as monsters of the deep stand

as sentinels at its portals, to contest our right to its navigation, we are forced to the conclusion that a branch line from
the Columbia River to Puget Sound must be established at
no distant date, even at the risk of a difficult two miles tunnel through the Cascade Mountains. In reference to our
connection with California direct, it is only a question of
time, and not one of cost or feasibility. When Sacramento
shall be connected with the Columbia by a line of 700 miles
of rail, a line so located will pass through a rich region of
the Sacramento Valley, and through the extensive mineral
districts of North California and South Oregon, and tap the
rich and large agricultural valley of the Willamette River,
where we have the largest bodies of agricultural land to be
found on the Pacific, where tracts of from 1,000 to 11,000
acres are under cultivation. One of the largest of these is
to be found in the Sacramento Valley at the estate of Major
Bidwell, where grain is reaped, not by the acre nor by the
mile, but by the league. His estate is in fact the beautiful
village of Chico, where in rural wealth and in rural simpli-
city live a happy and contented people, all more or less supported by the means of this bachelor millionaire, whose residence is one of those beautiful architectural gems hid away
amid shrubs, trees, orchards and forests, as if to avoid the
gaze of him whose residence is of crowded cities, and is almost unworthy to breathe the sweet perfume of a region
where such bowers grow. I do not know him, but may
such a noble almoner long live to dispense heaven's bounties to a people who love him for the liberal and generous
manner in which he shares his wealth with those not simi-
larly blessed.

To me the study of the geography and resources of the
North Pacific States and territories has been one of mingled

pleasure and satisfaction, and no one, be his profession what it may, can pass over this region without experiencing a thrill of pride in seeing what energy and capital have ac-complished during the past fourteen years of its occupancy; and even now to picture in imagination what the next fourteen years may produce, would almost lay me lia-ble to such a doubtful criticism that I forbear to enter upon a theme so pregnant with interest; suffice it to say, let those who have never traveled through this in-teresting region, make once at least the journey, if time and means be theirs, and see for themselves pleasant homes and well-tilled fields, grand mountains and useful rivers, forests of orchards and oceans of grain, miles of sluice-boxes and tons of gold, and the beauty of a region re-claimed from the wilderness, and now enlivened by the merry songs of industry, contentment, and thrift; or if they prefer to view a region not yet subjugated to the uses of man, or the sites of his happy home, let him traverse the Rocky Mountains, and see them clad in all their noble gran-deur. Where the primeval forest has never yet heard the woodman's axe; where the deer, the bear, the buffalo and elk, alone claim a residence; where nature has erected her temples far from the presence of man, and where natural re-ligion is breathed in every visible element of her solitudes; and if in the study of either picture here found, he is not well paid for all the toil and expense of his journey, the fault will certainly be his, and not the bounties of generous Nature, who, with a liberal, lavish hand, has spread so many pictures of the grand and beautiful in this distant region, nor yet the fault of the inhabitants by the wayside, who by cul-ture and improvement have framed these pictures in gilded casements, and where contentment and happiness are the

visible garments in which everything would seem to be en-
robed. Here the church and school-house have erected
their steeples, the barometers of the morality and intelligence of the people, and where warm, generous hearts bid
a welcome to the residents of a clime where civilization
boasts other, though no more worthy monuments. These
people have gone forth to plant your outposts on the Pacific,
and keep sentinel over your commerce that extends to the
Indies; they have gone forth amid difficulties and dangers
that have made many sturdy hearts hesitate, to there found
for you new, permanent and peaceful homes. Be to them
kind and generous, and extend to them the right hand of social and political charity; join them in their pride and
boast, that, as the Atlantic has given birth to New York, an
elder commercial emporium that looks towards the marts
of the Old World, the Pacific claims the parentage of
the younger New York in the golden city of San Francisco,
the commercial window of which looks out upon the ports
of China and Japan, which until to-day have been closed to
the world, and we shall ever cherish the hope that, as the elder and younger sisters, they may be seen linked in bonds
of peace, friendship, and commercial unity.

ADDENDA.

The original intention in the foregoing summary was
to publish in brief an account of the leading features of
the country drained by the Columbia and Missouri.
After the matter was put to press, I was induced by
many directly interested in the full development of
Idaho, Montana, Oregon, and Washington, to enlarge
the compass by including many important statistics to
date; and, in so doing, I have compiled from the best
information at my disposal such facts as we thought
pertinent, and in this connection I was kindly aided by
Robbins, Kerr, Creenzebeck and Wilson, of Idaho;
Isaacs and Walker, of Washington Territory; Bradford,
of Oregon; Chouteau, of St. Louis; Holbrook, Hussey,
Franklin, Bien, Leland, Tilton, Appleton, Holiday, and
Prof. A. K. Eaton, of New York; Judge Tufts, and Mr.
Purple, of Montana, and the agents of the Idaho and
Montana Mining Companies located in New York, to
all of whom I return my thanks. The extracts from the
leading journals have been compiled without much
reference to date or localities, but simply with a view of
presenting, in a condensed form, such material as related
to the boundless wealth of our Rocky Mountain chain.

It is now conceded that mining in its various departments is the paramount interest, not only in California,
but throughout the entire Pacific coast. In its several

branches of gold, silver, coal, and copper, it is entitled to our first consideration. Upon its successful prosecution and permanency more than upon that of any other pursuit our commercial, manufacturing, and agricultural prosperity depend. If it fails or languishes, these several branches of industry and sources of wealth must, in like manner, suffer. If it is neglected or permitted to decline, the growth of our cities must receive a check, real estate depreciate, population diminish, and every kind of business meet with a corresponding set-back. In these mining pursuits we have a field always affording remunerative employment to a large population, and ample to the profitable absorption of all our surplus labor and capital. The commodities they yield are not liable to accumulate on our hands, or suffer from competition and over-production. They always command their full value and find a ready and healthy market. It therefore behooves us, that, avoiding the excesses and correcting the mistakes of the past, we continue to uphold this right arm of our national growth and individual prosperity. If our past experience has been unfortunate, let us draw from it lessons of wisdom whereby greater success, if possible, may be achieved, and loss avoided in the future; but by no means determine to turn our backs upon and deny further aid to a calling that so completely underlies all our other material interests. That more caution should be observed in investing in, and better judgment displayed in the management of mines, is not only proper, but highly necessary. Let the same circumspection be used in engaging in this that is commonly exercised in other undertakings, and, with our present knowledge of the business, mining will be likely to prove

nearly as certain in its results as manufacturing and
mechanical, and even more safe than mercantile pursuits.
Already the great danger is past. Fancy-mining has had
its day. The spirit of speculation has exhausted itself.
Nearly everybody has consented to come down to
economy and hard work. Very few think now of selling wild-cat stocks, or even shares in unproductive mines.
The most persistent dealer in feet does not presume to
force this species of property longer on the public attention. A most sweeping and salutary change has taken
place in this particular. However hard up a man may
be, he no longer seeks to make a raise by selling this
class of stocks, or pledging them as collaterals. This sort
of mining, then, having had its day, bringing with it disappointment and disaster, will not, we may reasonably
believe, be soon reinstated. Among brokers and pro-
fessional speculators the old fictitious and inflated style
of doing things may, to some extent, be kept up; but
with business-men and practical miners we have reason
to know it has been about discarded. The latter, everywhere throughout the country, manifest a desire to have
done with it, and are anxious to work and develop their
claims before offering them for sale. This feeling is
almost universal with that class, and should be en-
couraged, at least to the extent of supplying them the
means to carry it out in practice. With our past experience, this pursuit promises to be much more successful than heretofore. The labors of many years have
taught our miners how to judge the value of ledges, as
well as the proper manner of opening them. Labor is
applied with better judgment, and expended with more
economy and effect, than before. When the various

mining localities have been tested—the poor districts
and the good have been found out—work will not so
frequently be misapplied or wholly thrown away as before; in consequence of all which, the expenditure will
be less, while the chances of striking something valuable
will be much greater now, than they were a year or two
since. Every one knows that while much money has
been lost, so also has much been made by working and
dealing in mines the past few years. To suppose these
opportunities are all gone, or greatly diminished, is a
mistaken notion. It is highly probable that even more
money will be made by shrewd and cautious men of
means in the coming than there has been during the past
five years. It could not be forced by the vehement
energy, nor be readily compassed by the mechanical ingenuity, of our people. It required patient experiment,
skilful manipulation, and the solution of the most in-
tricate chemical problems, and for a time baffled all our
efforts at success. Under these and manifold other difficulties and drawbacks was the pursuit of silver mining
installed on this coast; and if it has cost us some trouble
and sacrifice, we may console ourselves with the reflection that such has been the case with many other
branches of business, now well established and carried
on with profit; and with the further reflection that these
embarrassments and troubles are now pretty well over
with, and this great industrial interest placed on a firm
foundation. Our people, with characteristic skill and
perseverance, have succeeded in conquering the difficulties met with at the start, and are now not only
masters of the art, as they found it, but have advanced
it immeasurably since, by the introduction of improve-

ments, both chemical and mechanical, whereby great economies have been effected in all its departments. By repeated and careful experiments, combined with ingenuity and science, they have succeeded better than any other people in determining the question how to extract gold and silver from their matrix and associate metals, with profit; and it only now remains for us to prosecute this new and promising branch of business with energy and care; and benefiting by our dear-bought experience, avoid the follies that have heretofore been committed, and which alone have been the cause of so much loss and disappointment.

That mining upon this coast is not only a prosperous but a growing business, is shown by the large and increased product of bullion during the financial year just closed. The entire amount received at San Francisco amounts to $53,601,393, and surpassing that of any other year except one. This increase is due in part to the large amount reaching us from the new State of Nevada, very nearly, if not quite, $16,000,000 ; to liberal instalments from the northern gold-fields of Idaho, Washington, and Oregon; and, still more, to the very gratifying manner in which our own quartz-mills have contributed the past season—the amount received from this source being fully equal to, if not more, than that derived from Nevada. This business has been rapidly enlarging for the last few years, and being very generally successful, with an unlimited field open before it, we may look for its further if not rapid expansion. That the inducements for investing money in it are fully equal to any elsewhere presented, seems undeniable ; while as a home interest it has peculiar claims upon the support of our own people.

The following article, prepared by Mr. Walker, was intended to be published by him ; but on learning of my plans, kindly turned this and other material over to me, to be used at my discretion.

THE GOLD AND SILVER MINES OF IDAHO.

FUTURE developments will furnish material for abler pens to write a gold and silver history of the " Gem of the Mountains." At this early day a preface only to the future of her mines can be written.

The brief period which has elapsed since sufficient of the precious metals were known to exist, to warrant the working, and the difficulties with which a sparse mining population have had to contend ; with little capital to develop their interests, added to the persistency with which the press of San Francisco had discouraged emigration from her borders to these new gold-fields, has very greatly retarded the development of the rich and extensive placers, and the hundreds of gold and silver bearing quartz lodes, the value of which it is a part of our present purpose to establish beyond doubt or cavil by such evidence as we have been able to gather, not only from the local press of Oregon, Idaho, and Washington, and reliable correspondents in the mining region, added to our personal knowledge and observation, but the eleventh-hour testimony also of our ancient enemy, the press of the Golden City, which forced to do us justice at last, from the undeniable proofs presented at its very doors in the shape of huge bricks of bullion brought down from Idaho to Portland, Oregon, and shipped from that point by ocean-steamers, and which could not possibly have come from Washoe or Reese

River by that route, has very gracefully " caved in," and now acknowledges the mineral wealth of our Territory, and is making ample amends for its former seeming incredulity, by publishing established facts.

As Idaho is a *terra incognita* to the people of the Atlantic seaboard, it may be well to state briefly some few facts incident to the discovery of the precious metals in the Territory. This may date back to 1853, the year of the Northern Pacific Railroad Exploration and Survey, conducted by Governor (late General) Isaac I. Stevens. In traversing the vast region of country, then all embraced within the boundaries of the Territory of Washington, and lying between the Rocky Mountains and the waters of Puget Sound, and drained by the great Columbia and its tributaries, the exploring party discovered gold in the beds of small streams at various points.

A year later, the late General F. W. Lander, who was connected with the railroad surveys of Gov. Stevens, as chief civil engineer, conceived the idea of an exploration from the Dalles of the Columbia, striking diagonally across the lines of the main surveys, passing through the region of country between the 46° parallel and the southern boundary-line of the Territory, and entering the Rocky Mountain region, by crossing the Malade and Upper Snake River, and Sublett's cut-off between the " Old Crater " and the Wahsatch Mountains to Fort Bridger, for the purpose of examining the passes in that direction, with a view to the construction of a branch railroad from the California line of survey to the navigable waters of the Columbia. In making this reconnoissance, General Lander noticed the same indications of gold over the line of his march west of the Mountains, as were noticed by the party making the main surveys the year previous.

Several years later, Captain John Mullan, of the army, now a resident of Washington Territory, was directed by the Secretary of War to take charge of the construction of the military road from Fort Benton on the Missouri, to Fort Walla Walla, Washington Territory, a distance of six hundred and twenty-four miles. The construction of this work, passing as it did over the mountain region of what was then all Washington Territory, was naturally an opening examination of the large mining fields now lying in Idaho and Montana Territories, and which owe their chief discovery to Captain Mullan's explorations and researches, while constructing the longest continuous line of military road in the United States, from the head-waters of the Missouri, which drain the eastern side, to the navigable waters of the Columbia, which drain the western slope of the continent.

Soon after the completion of this work, steps were taken to more definitely ascertain whether the precious metals existed in sufficient quantities to pay for working, which fact was ascertained in October, 1861, when the

PLACER DIGGINGS OF THE TERRITORY

were discovered by Captain E. D. Pierce, an old Californian, but then an Indian trader, who had been for several years in what was then the northeastern portion of Washington Territory. He had found gold in several localities, and had attempted to work for it, but the Nez Perce Indians, owning the country, had forbidden any search for gold. At last, however, Captain Pierce succeeded in gaining the consent of the Indians, and with a small party commenced examinations which resulted in finding placers that would pay, at the mouth of the Oro Fino Creek, a branch of the Clearwater River. Thousands rushed to these new diggings in the spring of 1862, which resulted in the finding of rich placers on most of the tributaries of the Clearwater. Once the precious metal found, these adventurous spirits extended their explorations farther to the southward during the summer, and discovered the extensive placers of the Salmon River, sixty miles distant from the Clearwater.

Meantime a party from Oregon, led by a Mr. George Grimes, had crossed the Cascade and Blue Mountains, and pushing north had entered the mountain region of the Boise River, where, on the first of August, they discovered the famous placers of "Boise Basin," one hundred miles south of Salmon River, which have proved the most productive and probably the richest and most extensive placers on the American continent, outside of California. A few days after this discovery, the prospecting party were attacked by Indians, and the leader, Mr. Grimes, was killed. His companions buried him on the creek where they first discovered gold, and which is now known as "Grimes Creek."

The spring and summer of 1863 found the gold-hunters pushing their explorations for gold still farther to the south. A party of some thirty men, led by M. Jordan, leaving the "Boise Basin" trail where Boise City now stands, and crossing the Boise and Snake Rivers, soon struck a tributary of the Owyhee River, now known as "Jordan Creek." Here the party found rich placers, but the gold is not of as high value as from the other placers of the Territory, much of it being worth only

5

twelve dollars per ounce, having a large per-centage of silver. Jordan Creek flows through a narrow valley, flanked on either hand by broken ranges of mountains, known as the dividing ridge between the Great Snake or Lewis fork of the Columbia and the Owyhee Rivers, in which mountains have subsequently been discovered some of the richest and most extensive silver

QUARTZ MINES

on the continent. These mines may date their discovery from November, 1863 ; but not until February, 1864, were any special steps taken to further ascertain their extent; since which time over three hundred well-defined gold and silver quartz lodes have been discovered and located.

The outcroppings of these ledges contain a great per. centage of gold, but after being sunk upon, silver is found to predominate. This is the case particularly with ledges which, low down the mountain, come near the water level, as will be seen by the following assays of surface ores from two certain lodes named, the one, the " Oro Fino," being well up the mountain, while the " Morning Star" is some one thousand seven hundred feet below, near to, and crossing Jordan Creek.

"NEW YORK, Nov. 8, 1864.

" The sample of ore marked Oro Fino, Owyhee County, Idaho Ter., assayed, contains

<blockquote>
Gold to the value of $2016.24
And Silver   "   "   425.03
————$2441.27
</blockquote>

" The sample marked Morning Star, Owyhee County, Idaho Ter., contains

Gold to the value of $310.19
And Silver   "   "   2216.39
———————$2526.58

" Total values of each per ton of 2,000 lbs.

        " Very respectfully,

    (signed),   " EDWARD N. KENT, Chemist.

" No gold or silver visible in either sample before assay."

There are now three mills in the Owyhee district. The first one put up is known as the " Ainsworth" mill, and is the property of some gentlemen of the Oregon Steam Navigation Company, of which Captain J. C· Ainsworth, its late President, is one. The mill is of ten stamps, with every thing complete, and of superior workmanship, being built at one of the best machine-shops in San Francisco. It was at the mine in July last, and was intended to be run by water-power, being located on a fine bold stream; but on further consideration this plan was changed, and steam-power substituted instead, looking to an increase of the working capacity of the mill hereafter. This change involved the necessity of sending to San Francisco for engines, boilers, etc., which arrived on the ground last fall. Meantime the stamp went up, and was run with water-power, crushing considerable ore with the most satisfactory results. Operations were commenced on the company's mine in June last, which includes portions of the " Oro Fino," " Kirkpatrick," and " Navigation" ledges.

Work was commenced on the " Oro Fino," on which a shaft has been sunk over one hundred and twenty feet, and drifts run on the ledge. Several hundred tons of ore have been taken out, and the mill will be ready for vigorous operations early in the spring. For their machinery, opening the mines, and for other necessary improvements, the company have expended over one hundred thousand dollars in gold. Six months' run, however, with the ores they now have on hand, will put into their treasury this amount, besides paying expenses, and a handsome percentage besides.

The second mill erected, known as the "Moore and Fogus" mill, commenced operations about the first of November last, crushing rock taken from the "Oro Fino" and "Morning Star" lodes, in both of which this company own an interest. The ore on which they commenced work was that taken out all the way from the outcroppings to a depth of fifty feet, and the yield of gold and silver to the ton of this rock is the most extraordinary, probably, of any ores ever worked on the Pacific coast. Let it be borne in mind that, in the working results here given, there have been but the commonest means used; no smelting or roasting of ores, but simply amalgamation with quicksilver. What, then, will be the result when these ores are treated more properly and scientifically?

The third mill erected here belongs to Minear & Co., has only five stamps, and was put up for custom-work. It is now crushing rock from the "Allison," "Home Ticket," "Whiskey," and other ledges in the vicinity; and the yield from the rock worked at this mill, as well as the ores crushed at the "Moore and Fogus" mill, is shown by the following extracts from Idaho, Oregon, Washington, and San Francisco papers, dating from the first operations of these mills, to the latest advices received thence to date.

The extracts that here follow are of special value, and may be relied upon as the best authentic evidence now published :—

Daily Oregonian, Nov. 16, 1864.

OWYHEE STILL IMPROVING.—We are shown a private letter from Idaho City that relates some facts given by Major Holland on his return from Jordan Creek. Moore & Fogus' mill, known as the Morning Star mill, made its first eight days' run from the Oro Fino quartz, taken from the tunnel on the discovery claims, in which some of the mill-owners have large interests. This was only second quality of ore, used for the purpose of smoothing the mill, and the clean-up at the end of eight days' run showed 600 lbs. of amalgam after it was strained. This will give probably 400 lbs. of gold and silver; and as the Oro Fino is very rich in gold, it probably will yield $5 per ounce when run into bricks and correctly assayed. Even half of that would give $1,500 for each day's run, and the ledge at the discovery being six feet wide, the world can form its own estimate of the prospects of the fortunate owners.

The next run of the mill was ten tons of choice rock from the Morning Star ledge, in the discovery claim of which Messrs Moore & Fogus also have large interests, and the product of this was *two thousand pounds of amalgam*—ten per cent. of amalgam on the amount of rock crushed. The Morning Star contains less gold than the Oro Fino, but is immensely rich in silver, the rich vein being twenty-two inches wide, though the quartz is over three feet wide in some places. They expected to save some 500 pounds more of amalgam from the same rock, which at a moderate estimate would give $25,000 for perhaps two days' run. We gave yesterday particulars of what other mills are doing, and we see by the Boise *Statesman* that one mill will soon lay up until spring.

Daily Oregonian, Nov. 26.

BRICKS AND BULLION.—There was quite an arrival of bricks and bullion from the interior on Saturday night. Owyhee, Boise Basin, Kootenai, and John Day's mines, were each represented with their distinct features of bricks, unmixed, and the color was easily distinguishable. The Kootenai dust makes a very handsome brick, but the finest brick of all—the prettiest gold—was one worth $3,005 21, assayed on Saturday by Tracy & King, from the results of a crushing of ore out of the Ruckle vein of the Rockfellow lode. Had we the mint in operation here now, at least $1,000,000 worth of bullion would be deposited within the next twelve hours for coinage. As it is, there is not sufficient coin to buy the bullion offered, and bricks are exchanged for merchandise in numerous instances, to very large amounts. The city is crowded with merchants, miners, and traders, from the mineral districts.

TREASURE.—The treasure-box of Wells, Fargo & Co.'s Express, by the steamer *Wilson G. Hunt*, last evening, contained nearly $80,000 in gold and silver bullion. The season of mining has been closed for some time, but the flow of treasure does not seem to close with the season.

Oregonian, Nov. 28.

Mr. Marion Moore arrived here last night on the stage from Owyhee, with over 600 pounds of bullion, in seven bricks, from the Oro Fino and Morning Star ledges. They were obtained from 100 tons of rock. Mr. Walter Davis also had two small bricks, weighing together 28½ pounds, and valued at from $1,800 to $2,000.

Dalles (Oregon) Messenger, Nov. 11.

A GOLDEN HARVEST.—The owners of the "Oro Fino Ledge," Owyhee, have taken out and piled up alongside of their claim fifty tons of rock, which, at a moderate estimate, will yield $100,000. This is all picked rock, in which the gold can readily be detected by the eye, and is believed to be the richest rock ever taken out from a mining claim. This rock is about to be crushed, and in a brief space of time the bricks will be here on the way to the mint.

Daily Oregonian, Nov. 12.

NEWS FROM OWYHEE.—From Mr. G. C. Robbins, just returned from Owyhee, we learn that the mills are now steadily working, with a yield that is really astonishing, and fully equal to the expectations of all concerned. Crude bullion is being turned out *by the ton*, and we shall soon receive here the most substantial proof of the value of that region in the shape of the piles of silver bricks we have so often read of as being produced at Washoe. The Morning Star mill had been working rock from the Morning Star ledge, near which it stands, and the results were surprising; it being necessary to stop the pans frequently, as with eleven hundred pounds of quicksilver at work, the quicksilver soon became too thick to work to advantage, from the great accumulation of the real silver in the pan. Three tons of "Whiskey" rock yielded $1,000, being merely average rock. Minear & Co.'s mill was crushing Allison rock with good success, but would soon try the quartz from Home Ticket; when there is no doubt the well-established fame of that ledge will be built up on a solid foundation of "bricks."

The Ainsworth Mill Company were crushing good quartz at their mill, saving the gold, and are well paid for so doing, but not yet having the proper machinery to save the silver. They will soon be prepared in that respect, and then will have no difficulty in showing more sizeable bricks, though not so rich per ounce as at present.

The success of Owyhee is assured. There is probably no known silver-bearing district that has opened up more satisfactorily than this, and very few do so well.

The quartz as a general thing crushes easily, and the gold and silver is easily saved, so that many of the difficulties that retard the progress of other mining districts are unknown here.

While many of the ledges are of good size, there is no exception to the rule, that where a clearly-defined vein is found on the surface, it invariably improves in width and richness as it is sunk upon.

This is only the commencement of a rich and prosperous career of the mining region that lies on the head-waters of the Columbia River.

———

Dalles Mountaineer, Nov. 12.

HANDSOME BRICK.—Mr. Wm. C. Moody yesterday exhibited to us a brick from the Minear Mill, weighing one hundred and ninety-nine ounces, and valued at considerably over nine hundred dollars. The rock from which the bar was obtained was taken from the Whiskey Ledge. The brick assayed over a dollar to the ounce in silver, and some four dollars in gold.

———

San Francisco Bulletin, January 4.

SHIPMENTS OF SILVER FROM IDAHO.—The steamer *Sierra Nevada*, which arrived on Wednesday, brought down about $90,000 in silver bricks—the first considerable shipment of that metal from Owyhee, in southwestern Idaho. There are two qualities of bullion, one lot being worth $3 25 per ounce, and the other $6 40 per ounce. About four-fifths of the amount was of the former quality. One lot of about $57,000 is said to have been the product of fifty-five tons of ore, while another lot of about $33,000 is the result of fifteen tons' crushing. Some of the sanguine Idahoans are predicting that their young Territory is to excel Washoe as a silver-producing country. The facilities, such as wood and water, are said to be abundant in Owyhee.

Index, January 4.

A GOLD nugget, weighing forty ounces, was recently taken out of a claim on Grimes's Creek, Idaho Territory.

Bulletin, January 13.

An Oregon correspondent of the San Francisco *Bulletin*, relates the following of a silver-mining result in that State, worth, at $12 the ounce, $158,400:—

" A few days since, the largest pile of the precious metals my eyes ever beheld came down from the Owyhee country. The bricks, twelve in number, came from the Oro Fino and Morning Star ledges, and belonged to Moore & Fogus, who own the mill at which they were turned out. The metal rates as silver, though some of it contains sufficient gold to be valued at $12 per ounce. There was in all, I think, 1,100 pounds of that stuff. Seeing these great silver bars piled upon one another in the form of wild-turkey pens or log cabins, on the floor of the banking-house of Ladd & Tilton, made this metal, hitherto considered and denominated precious, appear as common as pig-iron.

Daily Oregonian, December 13.

INDICATIONS.—We have been telling the world, for some time past, wondrous stories of the wealth of Owyhee. The many ledges and great richness of them are themes all the readers of our paper have been conversant with ever since they were discovered, for we have been among the mountain-crags on Jordan Creek, spent months in clambering over hills that bear riches equal to those of Arabian tales, could the wand of some friendly

5*

genii only reveal them to the sight. And it required more than the ordinary amount of self-control to wander among the laboring prospectors and not become over-enthusiastic at the "prospects" that were brought to view by their patient toil. As for the ledges, "their name is legion," and there is room for working-legions to develop them, and practical proof of this is to be seen now by all. Marion Moore, Esq., one of the principal owners in the Oro Fino and Morning Star ledges, has lately come down with over $60,000 in silver bricks, duly impregnated with gold. The largest of these is something smaller than a soap-box, and the smallest will tax one's sinews somewhat to lift. Then it is the genuine and indisputable product of the two Owyhee ledges above named, and hundreds have seen it, and the rest can see it by looking into the banking-rooms of Messrs. Ladd & Tilton.

But while you are looking at the genuine "*argent,*" don't forget to see and handle the black-looking specimen that lies near them, which is not exactly quartz, but is something better, being a piece of ore from the Morning Star Discovery Shaft, known as "*polybasite,*" weighing some six pounds, we should judge, and assaying nearly seventy-five per cent. of silver, falling not far short of the melted bars in richness.

Oregon Advocate, December 17.

LARGE SHIPMENT OF TREASURE.—The *Sierra Nevada,* which sailed from Portland on the 2d inst., carried treasure to the amount of $517,250, shipped by four firms of Portland—Wells, Fargo & Co. sending $450,000 of the amount. The *Oregonian* says a large sum besides went in the hands of passengers.

Dalles Mountaineer, December 20.

SILVER BRICKS.—Mr. Grenzebach arrived at the Dalles by yesterday's boat, bringing with him a number of silver bricks, the aggregate value of which was $3,000. These bricks were from the Minear Mill, and, taken in connection with other bricks from this mill, would seem to indicate that the milling as well as mining interests are prosperous in the Owyhee country.

Daily Oregonian, December 13.

TREASURE-DRIFT.—The treasure-drift indicates to a certain extent the prosperity of our miners, and is kept up in a steady and constant flow, with occasional strikes, such as appeared Saturday night, for instance, on the arrival of the steamer *Wilson G. Hunt*, from the Cascades. Mr. West (Wells, Fargo & Co.'s messenger) had in charge that evening, *over eight hundred pounds in bullion.* Passengers by the same boat brought an aggregate amount equally as large. The demand for coin for the purchase of dust is at present very active. Among the shipments of treasure to-morrow will be a large amount from the Kootenai.

ASSAYS.

From the Oregon Daily Times, November 28.

" The news from the Owyhee mining district is sufficient to warrant the broad assertion that Oregon and Idaho Territory will soon be recognized as the richest countries in mineral wealth on the face of the globe. Mr. Luther Hasbrook, a gentleman who has just returned from the Owyhee, called on us on Thursday last, and exhibited to us a sack full of quartz specimens, taken mostly from the Oro Fino, Morning Star, Evening Star, and Noonday lodes. Contrary to the usual custom of selecting the richest specimens for assay, Mr. H. brought down with him every character of rock which had been obtained in the lode, from the richest to the poorest. A chemical analysis of that taken from the Oro Fino shows a valuation of $22,000 per ton, with the proportion of $3,000 per ton in silver. From the Morning Star, an average of $11,000 in gold and $2,000 in silver has been taken from a ton of the rock. The Evening Star is said to equal the Morning Star, while the Noonday, it is believed, will surpass all the others in richness."

Extraordinary Rich Silver Ore.

"We were present, on Saturday last, in the laboratory of Professor Blake, on Montgomery street, when he was making an analysis of several specimens of ores sent from the Boise region for his examination. Among these specimens was one that, to all appearances, was what a good mineralogist would pronounce at a glance to be copper—a very rich ore of copper —but which, on analysis, proved to be the most extraordinary ore of silver probably ever found in the world. It yielded, in a great number of tests, the great proportion of eighty-six per cent. of silver and some gold, being of the estimated value of twenty-six thousand dollars per ton! From the specimen operated upon, we should judge that it came from a large mass of the same kind of ore, which is a unique variety of massive sulphuret of silver. On a thorough analysis, it was found to contain traces of arsenic, sulphur, antimony, and copper, the residue being silver. Besides this extraordinary specimen, there were several others containing a large per-centage of chloride and native iodide of silver, and sulphurets of that metal in crystals, very beautiful and very rare, and very valuable, too. If the Boise region contains much of such ore, Washoe will be nowhere, and the Comstock ledge a mere bagatelle."

Results of a Quartz Assay.

"Our readers will remember the departure of Mr. Donovan for San Francisco recently, with 1,500 lbs. of rock from the Home Ticket lode in the Owyhee mining region, and they will also remember our allusions to it as being perhaps the richest lot of quartz ever taken from this section. Our statements are corroborated by a private telegram received from San Francisco, giving the results of an assay of that rock made by Messrs. Molitors, Assayers, showing it to yield in

Silver to the ton	. $6,707 61
Gold "	. 489 12
Total value, per ton	. $7,196 73

which is nearly $1,000 more than the richest assay ever made of the celebrated Gould and Curry of Washoe notoriety. One ton of the rock of this noted claim was culled over and sent to Europe, which resulted in value to the sum of about $6,560. The rock taken below by Mr. Donovan was procured at the depth of about eighteen feet, on the Discovery claim of the Home Ticket lode, and is only average rock, as we understand from reliable sources. Our citizens are deeply interested in the development of these mines; and all such evidences of their superiority to the renowned Washoe, will be received with gladness. An assayer now in the Owyhee region, from Reese River, gives it as his opinion that this country has the richest rock and the greatest amount of gold.

GENERAL ITEMS.

BOISE BASIN DISTRICT.

[FROM AN OCCASIONAL CORRESPONDENT.]

WALLA WALLA, W. T., December 8, 1864.

My last advices from Boise say that the people there have gone into spasms under the influence of quartz developments. The fever is likely to become an epidemic in that whole region of country. My correspondent writes that the Gambrinus Company's mill, near Bannock City, cleaned up in five days' run, $15,000. The stock was selling at $100 dollars per foot.

Late discoveries at South Boise of gold and silver leads have created considerable excitement, and predictions are common that an area of 100 miles of the present discoveries, including the Yuba and Volcano districts, and the Owyhee, will prove the richest quartz mines which have been found on this coast. In conversation with an intelligent gentleman, who has made explorations in that country during the past summer, he confirms the above opinion. Two quartz-mills are now in operation at South Boise district. Another year will determine whether this northern country is of sufficient mineral importance to enlist capital on a similar scale to the Washoe country.

Explorations are being made on the head-waters of the Columbia, and reports come in of valuable discoveries.

The Gambrinus Mill

Is giving a good account of itself. The second clean-up, after a run of three days, gave about there hundred pounds of amalgam. Stock is now held at a high figure, and the holders thereof are exceedingly jubilant. The ledge is undoubtedly one of the richest ever worked in this or any other part of the United States. The gratifying result of the crushing is giving surrounding ledges a new impetus, and has a very encouraging effect among mineral owners generally. There is no doubt that many other ledges, equally as rich, will soon show their true worth under the unappealing despotism of the iron stamps preparing for them.

South Boise.

Mr. Comstock.—This gentleman, whose experience in quartz has been great, and who has been a practisal miner in all the branches practised in New Mexico, California, Nevada, Oregon, and Idaho, for sixteen years past, has been giving us some items relative to the character and importance of the mining districts lying on the head-waters of the Columbia River.

Speaking of the quartz ledges found in the Boise Basin and in South Boise, in which he has lately been much interested, and where he has done much to develop and prospect the country, he assures us that he considers, with the light of all his past experience to aid him, that country exceeds, in the extent and richness of its gold and silver bearing quartz, any thing that he has ever known in life. And when we recollect that he gave his name to the celebrated Washoe ledge, and has been ever since searching the mountains for similar developments, we must acknowledge that his opinion is entitled to certain weight and importance not due to all others.

We gave a few days since a sketch of the quartz ledges in South Boise, which Comstock fully indorses, although he insists, that on all points, we are far below the truth, if erring in any respect. We hear of many mills destined to be erected in the eastern quartz mines another year, and have no doubt of the success that awaits the investment of capital there, and cordially hope that the hardy and enterprising pioneers, who have opened up these treasures to the world, may reap their full proportion of the harvest that is in store.

ITEMS FROM SOUTH BOISE, I. T.—From J. W. McBride, of Alturas County, or what is better known as the South Boise country, I. T., the *Oregonian* of December 6th gathers the following items:—

The quartz lodes discovered in this county are hundreds in number, and enough of them have been prospected and worked to render it certain that this will rival any portion of the Pacific coast in the richness of its mines and in the extent of the same, so soon as mills can be erected in sufficient number to offer the proprietors of ledges opportunity to test and work them. One five-stamp mill, owned by Carter & Co., has been at work during part of the summer and fall, and has crushed about five tons of rock per day, with good results. It had crushed some one hundred and fifty tons of Ada Elmore rock, with an average result of something near one hundred dollars per ton. It is now engaged in crushing rock from the Confederate Star ledge, which the owners thought would go as high as one hundred and fifty dollars per ton. For some time past arastras had been at work upon a number of ledges in this the Bear Creek District, and a substantial building was erected for a mill, running six arastras by water-power, and every ledge had so far shown good results upon being worked. The rock was much of it rich in silver—one assay had shown two hundred and twenty-five dollars in silver and eight hundred dollars in gold per ton—but no efforts had yet been made to save the silver, though the mills would soon be prepared with the best apparatus to save both silver and gold. Waddingham and McBride have fourteen thousand pounds shipped on the way down, which they intend to have throughly tested, sending part of it to Swansea, Wales, if necessary, to secure a proper working and a thorough analysis of their ore, and so be prepared to erect works to reduce it in the best manner at the mines.

A larger mill had arrived at Rocky Bar overland from Chicago, working twelve stamps, which will be running this month; and a Mr. Fox was going in with another five-stamp mill, to be erected in the same district.

Bear Creek District is some eighty miles northeast from Boise City, and fifty miles southeast of Idaho City. Rocky Bar is a thriving place, with stores, hotels, etc. Some three hundred persons will winter there, and some six hundred in that camp. There were two thousand miners in that country during the summer, and there will be as many another year.

The ledges are found in granite formation, in each district. Yuba District is thirty miles northeast of Rocky Bar. Comstock lately struck a good ledge here, eight feet wide, and rich in silver and gold.

Volcano is thirty miles southeast of Rocky Bar, where there are good ledges and rich placer mines. Ledges are well defined and of solid quartz all through the country, average three feet wide, and some much wider; not so rich as a few Owyhee ledges, but perhaps larger in size, and able to furnish a reliable amount of good-paying quartz.

SOUTH BOISE.—Mr. Wilson Waddingham arrived last evening from Rocky Bar, Alturas County, Idaho Territory, bringing with him some fourteen thousand pounds of quartz from different ledges there, and also some solid results of the value of the ore in the shape of "bricks." One is worth about one thousand dollars, and several others range from two hundred dollars to seven hundred dollars, each of which represents a different ledge, showing that the average yield is over one hundred dollars per ton, without saving the silver which exists largely in the ore. The gold is valued at twelve dollars per ounce, and the whole is a very handsome foreshadowing of a good time coming for South Boise. We also learn of the arrival last evening of a large quantity of silver bullion from Owyhee, in care of Mr. Moore.

Quartz ledges continue to be discovered and located, the richness and extent of which must remain a question in some degree until another season at least. Two quartz-mills are in process of erection within a few miles of this city, one owned by R. C. Coombs & Co., upon the Gambrinus Ledge; and the other by Williams, agent of the Boise River Gold and Silver Consolidated Co., upon the Landon Ledge, about a mile distant from the Gambrinus. The mill of Raymond & Co., near Placerville, has been in successful operation for some time past. So the quartz interests of the country, as I predicted in a communication to you about a year ago, are at last in a fair way of being developed fairly and fully. It is not to be inferred, however, that one hundred quartz-mills, if the machinery were now in this country (including Owyhee, South Boise, North Boise, Grimes Creek, and the ledges in the vicinity respectively of Placerville, Centreville, Pioneer, and Idaho City), would more than experiment a little upon the astonishingly rich and extensive quartz-bearing mineral resources of this region for the next five years. This may seem exaggerative, but the experience of after-years will prove its truth. Indeed, it is clearly apparent to many here now. Next year some income will be derived by the mill-owners of, and owners of rock crushed by, the few mills in operation, but the entire business is as yet in its infancy. Whether twenty-one years will bring it to full manhood is exceedingly doubtful.

[From the Daily of Wednesday, September 28.]

THE WEALTH OF THE MOUNTAINS.—We are favored with a private letter from one of the old residents of this city, which contains very much of interest relating to the prospects of the upper Columbia country, and we think we cannot devote the same space to matters of greater importance to-day, hence we extract. The writer, General McCarver, says of the newly-discovered quartz region : "The Silver Hill district was discovered about July 3d, by accident, as I learn from one of the party. They were attempting to find a new route from the South Boise mines to Placerville, by keeping near the Payette range of mountains, but finding the cañon near the base of the mountains impassable, determined to retrace their steps, and pass *via* Moore's Creek and Idaho City. On emerging from the cañon to the summit of a high ridge—now Silver Hill—they discovered the croppings of extensive ledges of gold and silver bearing quartz. A few specimens were taken to their cabin near Placerville, where they remained for a month without being tested—so little was their knowledge of the great wealth they had staggered upon in their rambles through the mountains —but finally a test with borax, in a blacksmith shop, developed the truth of its riches, and set the camps aglow for feet. When I reached Silver Hill I found about one hundred miners in and about two newly-laid-out towns—Banner City and Eureka— and strange as it may seem, some of the elderly gentlemen remark that they can recollect when there was not a house to be seen in either of the places! Lots are being pre-empted rapidly, and some day we expect to rival the value of front lots in Portland. There are perhaps fifty ledges, bearing different names, already discovered, and more are being struck every day. They are generally from two to eight feet in width. The Banner was the first, and I believe is the best, yet struck. It can be easily traced for several miles by its croppings. The California, the Washoe, the Black Warrior, and the Silver Age, are astonishingly rich in silver. We are now in the midst of a quartz 'feet' excitement—such as was hardly surpassed in the palmy days of Washoe—and when I tell you that we have richer, and more extensive quartz mines in this region than has ever been known or found, in California or Washoe, or both together, I assert that which hundreds of men in the southern country will scoff at, but which assertion history will nevertheless sustain me in."

The regular correspondent of the *Alta Californian*, writing from these mines in July last, presents the following statistics concerning their product :—

" Our quartz-mills are going up rapidly. Some have already commenced crushing, and, as I have before written and again repeat, the first yield from this interest will confirm all that has been said in their favor, and no doubt is entertained but that a dividend will be declared before Christmas.

" Gold is now coming down freely, and I am well informed that the exports of treasure during August will exceed those of any previous month ; a large proportion will, as heretofore, pass through in private hands, but a large increase in manifest shipments will be apparent. During the months of June and July, when trade was stagnant and mercantile remittances light, the following amounts of gold dust were assayed by the two principal offices :—

Wells, Fargo, & Co..........	$559,375 19
Tracey & King.................................	575,818 75
Total......................	$1,185,193 94

"Not being permitted to publicly use the return of Gold-smith Bros., I cannot honorably add to the amount; but as I am cognizant they are doing a fair business, it is presumed the sum total will reach one and one-half millions of dollars. During the same period, I selected the names of a few prominent firms as the purchasers of gold in dust and bars:—

Ladd & Tilton, bars.........	$191,365
Allen & Lewis, bars	143,732
Fuller & Co., dust	118,125
Tracey & King, dust........	112,362
Wells, Fargo, & Co., bars and dust	558,455
Showing a total of	$1,119,089

" Although these firms are the principal purchasers of gold, there are many firms not enumerated, who buy and ship various amounts, which in the aggregate sum up a high figure. The information herein given is not guessed at, but, at some trouble, collected from the ledgers of honorable business-men, who have not a single object in view by this publication. After diligent, and not hap-hazard calculation, I have no hesitation in stating that the above is about one-third of what comes from the mines to Portland, independent of what goes overland from the mining country (which is considerable) to Salt Lake and California, besides what is being invested in farming, agricultural, and grazing operations in Idaho and Montana Territories. That the open shipments, as per manifest, will, during the ensuing months, average one million and a quarter per month, I firmly believe."

The Indian difficulties, which have so seriously interfered with mining operations in our new Territories, are now set at rest in Idaho. We copy the following from the Idaho *Statesman* :—

(*By Overland Telegraph to the Associated Press of New York.*)

SAN FRANCISCO, *March* 1, 1864.

" The steamer Pacific left here to-day for the northern coast with over a thousand passengers, mostly bound for the Idaho mines. Emigration in that direction overland has also commenced on a large scale, considering the earliness of the season."

SAN FRANCISCO, *March* 30, 1864.

"The rush of emigration from Oregon and California to Idaho is immense."

The present product of these mines may be inferred from the following telegraphic dispatches from San Francisco :—

SAN FRANCISCO, *Nov.* 5, 1864.

"The receipts of bullion from the interior and northern coast during the past ten days have been unusually large, and exceed *twenty-three hundred thousand dollars.*

"This is a gratifying exhibit, and shows the unabated productiveness and value of the leading interest of the people of this coast."

SAN FRANCISCO, *Nov.* 14, 1864.

"The Sierra Nevada came in from the north, Saturday evening, bringing $460,073 in treasure from the northern mines."

SAN FRANCISCO, *Nov.* 14, 1864.

"The steamer Pacific arrived here yesterday from the northern ports, bringing $548,501 in treasure—of which $260,375 came from Victoria, and $291,826 from Portland and Boise River."

SAN FRANCISCO, *Wednesday, Dec.* 7, 1864.

"The steamer from Portland, Oregon, brings nearly half a million in gold from the northern mines."

SAN FRANCISCO, *Monday, Dec.* 12, 1864.

"The treasure receipts from the interior for the last ten days are nearly two millions and a quarter."

SAN FRANCISCO, *Dec.* 14, 1864.

"The Oregon steamer, Brother Jonathan, brings $400,200 in gold, besides $150,000 in the hands of passengers "

SAN FRANCISCO, *Dec.* 31, 1864.

"Sixty thousand dollars in silver bars have been received at San Francisco from Idaho, the first shipment of silver from the new mines in southwestern Idaho, which promise to rival those in Nevada."

THE LANDON AND GAMBRINUS.

These two ledges are gradually gaining popularity as the prospecting advances. The former company have sunk on their ledge about thirty feet; but owing to the delay of the machinery which they have contracted for, they have stopped operations in their shaft for the present, and are working rock from near the surface for convenience. They are working three pestles with spring-pole hammers, and by this rude method are averaging from selected rock from ten to twenty dollars per day to the hand, crushing from one to two hundred pounds each. The Gambrinus, which in our estimation ranks with the best in the Basin, is rapidly putting things in shape for their mill, which is to be erected by Combs & Co. at an early day. This ledge is situated upon a side-hill, and stands dipping to the south at an angle of about sixty-five degrees, and almost parallel with the surface of the hill. The peculiarity of this location renders it an easy matter to lay a great deal of the ledge bare, by merely ground-sluicing the surface off. They have already stripped and in sight probably five hundred tons, not a particle of which will not prospect well. Some five or six tons of this rock was worked by an arastra the latter part of Spring, which yielded from three to eight hundred dollars per ton.

THE QUARTZ INTEREST OF IDAHO TERRITORY.

A six months' tour through the mining region of Boise Basin has afforded me favorable advantages for becoming acquainted with, and forming an opinion in relation to the future prospects of the upper country; and my acquaintance with the different mining localities, authorizes me in stating that the quartz mines will shortly become the paramount interest of Idaho. Although in their infancy, and in a very crude state, quartz mining is assuming a position that, with a little encouragement, bids fair to attract the attention of both capitalists and speculators. The recent discovery of several rich gold and silver lodes has given an increased impetus to further prospecting, and will ultimately become a source of wealth to those who can grasp the golden opportunity, and take time by the forelock. All that is required to develop their value is the introduction of machinery. There are now ninety distinct lodes recorded in the county office, on each of which from eight to ten thousand feet have been taken up and traced by the different claimants; each man is allowed two hundred feet by claim, and as much more as he likes to purchase. In several of these lodes men are making good wages with a common arastra; and in the immediate vicinity the dry-gulch mines are, or rather were when they had water, taking out from ten to twelve dollars per day to the hand, with a single rocker. Without pretending to a knowledge of which is the richest, I will confine my remarks to some of those which have more particularly come under my notice.

In Placer District, the three principal lodes (and I think not in the least varying in their richness) are the Pioneer (in which G. Collier Robins, of Portland, is a large shareholder), Golden Gate, and Lawyer; on the former of these several tons have bee. taken out, and an average assay shows $250 to the ton. They are situated in the immediate vicinity of Granite Creek, and, with the exception of a small gulch between them, are a continuation of each other. On the same range of hills, continuing a southeast course, the slope meets Grimes's Creek, where several lodes have been discovered, the one best authenticated being the Everest lode, above Hogem. The range continues through a spotted mining country until **we** reach El Dorado Gulch; here, for several miles, are rich placer diggings, and in the vicinity are the Mammoth Landon and Guymas lodes—from this, for a distance of fifteen miles, running parallel with Elk Creek, is one continuous range of quartz. Over one thousand claims have been recorded in this district alone, not mythical ones, but on each of which gold and silver in more or less quantities are found.

From an inferior piece of ore taken indiscriminately from a pile of the Mammoth, in which gold was entirely invisible, Messrs. Tracy & King, of Portland, make the following return:

One ton of rock yields twenty-four ounces of gold, value per ounce $16 50, fineness 800; total yield per ton, $450. The bars of this and the other Portland assay-offices rate on an equality with the best assays of San Francisco, and their established reputation is a sufficient guarantee for the correctness of the result. Steam—steam is all that is wanted to extract millions from the crumbling rocks. Most of the quartz is of a soft, porous, calcined nature; descending a few feet, it changes to a bluish crystallized appearance, sparkling with mica and pyrites; but in most of the lodes which have been traced, decomposed, rotten quartz prevails. As far as the Mammoth and Landon have been traced, ore has been found exceeding the most sanguine expectations.   CHAOS.

HOW IT WAS DISCOVERED.

The following is the history of the discovery of Rockfellow's new quartz lode, on the Boise road, which has its interest, as all such things have. Two men (names not recollected) were going to Boise; one of them had never been in the mines, and had never seen any quartz. As they were riding along, he mentioned this fact, when his companion, observing some pieces lying by the road-side, told him that was quartz, and they got off and picked up some of the pieces. After carrying them along some distance they found, upon close inspection, that they contained gold, and they took the pieces into Bannock

City. On their arrival there, their horses were stolen, and they left without means. They went into Rockfellow & Co.'s express-office, and while there, exhibited their quartz specimens. Mr. R. examined the specimens, and purchased one of the men a horse, and brought him back as a guide to show him where he found the quartz. On their arrival upon the ground, they found more of it, and traced it to the lode without difficulty, some two miles up the hill-side. A brief examination of the lode proved to them that, in miners' parlance, they had "found it," and the unfortunate, houseless, moneyless man, who went into the express-office two weeks ago with his quartz specimens, little dreaming of what was in store for him, is to-day the owner of an interest in probably one of the richest quartz lodes ever found on the Pacific coast—a mine of wealth equal to tens of thousands of dollars.— *Walla Walla Statesman.*

A TRIP TO THE ASTOR LEDGE.

On Sunday last, in company with the judge of the district court, district-attorney, sheriff, auditor, a member of the bar, an assayer, and one of the owners in the mine, we visited the celebrated Astor ledge, some six miles up Moore's Creek. Our route lay along Bear Run as far as the saw-mill; thence to the right across a ridge by a good wagon-road some four and a half miles farther to the ice-house; from here to the Discovery prospect, about a mile, the trail is good, but wagons have gone no farther up the creek than the ice-house. Arriving at the lead without difficulty, we all alighted and began our scientific researches after rich quartz. The strictest scrutiny on our part—being a novice in the business—failed to detect the existence of a single particle of metal; being assured, however, of the vastness of the wealth of the mine, and of the valuable character of the rock that lay loose around us, we determined to secure a piece broken from the ledge with our own hands,

6

for the purpose of testing it thoroughly, as the balance of the party had already resolved to do. As there was but one hammer in the party, we took turns, and each descended into the shaft and made his own selection—if it can be called a selection to break off rock from a ledge four feet wide, in no portion of which is visible a single speck of any thing valuable. With these—after tracing the ledge two or three thousand feet—we descended the hill to camp near a clear running stream, and refreshed ourselves with pure cold water(?) Here we procured implements and improvised a miniature quartz-mill, crushing a portion of the rock procured from the shaft on the hill, and washing it with an iron pan. The result of our experiments proved conclusively to our minds that the Astor is among one of the rich ledges. Out of about a spoonful of pulp, in each instance something near a cent in gold was obtained. The company has a tunnel run into the hill about fifty feet, which will in a few feet farther strike the ledge about one hundred feet below the surface. The facilities for working this mine will be extraordinary; a ravine some one hundred and fifty or two hundred feet below the tunnel, up which a most excellent wagon-road can be constructed at a trifling expense, will enable them to run in and drain perfectly, so as to strike the ledge from two hundred to three hundred feet from the surface. On Monday, Mr. Heald made an assay of rock taken from three different places in the ledge at random, by the parties who visited it the day before, with the following results: Gold, per ton, at $16 to the ounce, $453 93. Mr. H., the assayer, also procured some rock from the bottom, top, and edge of the excavation, from which by assay he obtained $240 09 to the ton in gold. Rock selected by Judge Smith, assayed by Rossi & Robie, $653 40 in gold, and $10 57 in silver; total, $663 97.

FROM OWYHEE.

Ruby City, June 1, 1864.

Editor News:—

Confined within doors to-day by a severe rain-storm, I thought I would pen a few lines to the *News*, as a kind of general letter to those residing at Boise Basin, having a holding interest here, with a slight digression of things in general, both prospective and otherwise.

PLACER MINES.

The placer mines in this section of country are more extensive than supposed. Jordan Creek, the main creek in this quarter, is worked for a distance of about eighteen miles; it is as large as Elk Creek, and many of the claims pay from fifty to one hundred dollars to the hand. The entire length of the creek is taken up. There are also several gulches which pay very well, and the hills in the vicinity prospect well, but water is difficult to be got on them. The placer diggings are not confined to this particular quarter; about seven miles south of here, placer diggings exist which will be worked before long, the only difficulty being a scarcity of water.

QUARTZ MINING.

To commence upon this topic, I may say, with a great deal of confidence, that this camp promises to become another Washoe. The number, size, and richness of the ledges, are unequalled, and work is being prosecuted upon many of them with great diligence. Among the principal ledges which are known to be rich, I may make mention of the following number, being immediately in this vicinity: War Eagle, Oro Fino, Morning Star, Alabama, Donelly, Silver Legion, Home Ticket, Allison, Whiskey, Empire, Eclipse, First Consul, General Grant, General Banks, Oxford, and others of which I know but little except from information. Moore & Co. are diligently at work finishing the wood-work of their mill; Crane & Co. are busily employed in boring through Oro Fino, or Silver Mountain, with much success. There is quite a large quantity of rock thrown up from the numerous shafts sunk on the above leads, and only wait the coming of machinery to "ship bullion." Quite an excitement occurred here of late, by the discovery of another rich quartz district, about nine miles from Ruby City, on a direct line south. The leads are silver, and promise to be very rich. The principal leads yet discovered in the new districts are—Monitor, Twilight, North American, Great Eastern, Siskiyou, Rising Sun, Spray, Setting Sun, Peace Democrat, Pride of the West, Gem, Blue Lead, Pyramid, and Rising Star. The rock from several of the ledges has been tested, and assays largely. The Mammoth District, about forty-five miles from here, is going to create an excitement in the world of San Francisco ere long. The ledges are immensely large, and from numerous assays made of the rock at Bannock, Portland, and San Francisco, the average top rock beats the Gould & Curry, of Washoe fame. The principal ledges in that district are— Mammoth, Occidental, Union, Bluster, Salamander, Occident, Diadem, Banner, Placerville, Wisconsin, Deer, Orient, and Ironclad. These ledges are from three to twenty feet in width, and can be traced a distance of from one to two thousand feet above the ground; so, also, of the lodes in the newly-discovered district. Who says this is not a great country? Why, in a year from now, from fifty to one hundred thousand men can find employment upon quartz ledges alone; and if the Government should tax the quartz miner as proposed, the snug little sum of about two millions of dollars must be handed in annually from these districts alone; that is, provided it can be collected.

ROADS.

Five parties are now busily at work in constructing wagon-roads from this place to Red Bluff, and from this place to Idaho City. The distance, they tell me, is as follows: From Red Bluff to

Ruby City, three hundred and fifty miles; to Idaho City, four hundred and fifty miles; but they think that the distance may be shortened at least fifty miles. The road from here crosses Owyhee River near the junction of Jordan Creek, and runs thence to Red Bluff; it will also shorten the route to Idaho City some ten miles.

Mr. D. P. Barnes, and a party of some thirty-eight others, arrived here last Saturday evening, after an absence of some two months' prospecting on Raft River, Goose Creek, and other localities in the direction of Salt Lake. They found some gold in different places, but nothing very rich. Copper, they say, exists in large quantities in that region, as well as immense quantities of marble of various colors, some variegated and beautiful, but so far from transportation that it will remain valueless until railroads are constructed to take it to market.

Yours, &c.,   OWYHEE.

NEVADA MINE-SALES EAST.

If the speculation in undeveloped mines which prevailed last year had continued to this time, countless millions would have been invested through companies incorporated in the East. The Pacific State operators in mining-stocks should have gone earlier to the Atlantic States, to have commanded the confidence of the general stock-market there, and set the people in excitement. However, the fall of stock-rates has not entirely prevented the growth of Eastern confidence in the richness of the distant mines. Several companies with large capital—stated, of course, to be "working capital," or "means to immediately develop the mine"—have, since the depression commenced here, been formed in New York and Europe. Some of the discoverers of, or rather the locators of claims in the silver-mines of this State, report that they have recently sold out to Eastern companies on magnificent terms. The Virginia *Union* assumes to have reliable information that a mine in the recently-discovered Pinewood District, Humboldt County, has been incorporated in New York City, and disposed of for the sum of three million dollars. It is the Pine Mountain Consolidated Mine, and comprises four ledges, of one thousand five hundred feet each, making six thousand feet in the entire claim. The Government stamps on the deed of transfer cost three thousand dollars, probably the largest amount of stamps ever placed upon a single instrument. This deed is now on its way from New York, for record in Humboldt County. The names of the locators of this mine are G. M. Marshall, William M. Hurst, better known as Sheba Hurst, and Amos H. Rose. They left San Francisco for New York about the first of last July, taking with them several tons of rock from the mine, the richness of which created quite a furor in Wall street. A working capital of one million dollars has been paid into the treasury, and the stock issued is all full paid and unassessable. The Board of Trustees, which is composed of several well-known capitalists in New York City, have already purchased machinery for a first-class quartz-mill, which will be shipped to San Francisco immediately, and erected at the mine as soon as possible. The large amount of capital already secured will develop the mine rapidly; and, with the abundance of wood and water in the district, the company hope to pay dividends shortly after the machinery arrives and is put up, which is not difficult to do where rich ore is found, and there are no assessments whatever.—*Washoe Star, Nov.* 12.

BOISE MINES.

POMONA, I. T., Sept. 8, 1864.

EDITOR OREGONIAN:—

The dreaded winter is at hand, and general work has made its exit. Gloom without; snowflakes adding a mist of bewilderment among the tall pines and craggy mountains. When the train of thought leaves the miserable *without*, and returns within our door, the scenery changes, and we recall by-gone days. Before going any further, let me describe our cabin, which is one-half mile from Centreville. To the left of the stage-road leading to Idaho City, stands a log cabin, ten by twelve feet, the roof extending eight feet from the main building, a pile of pitch-wood to the left of the door; over the wood hangs the fore and hind quarters of a beef. Under the same porch is seen a hand-sleigh, used for sledding wood and articles from town. We open the door and go in. Description is almost impossible, but I will endeavor to depict the scene. On the left side of the room is stored any amount of provisions, over which are fixed two bunks, one above the other. To the right of the fireplace stands a small table, on which are piled books, papers, and many other small articles too numerous to mention; and still to the right is a goods-box nailed to the wall for a cupboard, which is filled with all kinds of cooking-traps. On the right-hand side of the room is the window; one pane of glass constitutes the size, under which is placed the dining-table. The right-hand side of the room is ornamented with a large mirror and pictures; among them are seen Abraham Lincoln and his secretaries, generals, forts, battles, etc.

Peace and stillness reign supreme. Memory cannot be contented. The past is viewed as imperfectly spent, the future pictured in glowing colors, and a world of wealth and happiness within a year's grasp. Our mountain home—three years ago a wild, savage wilderness—has undergone a serious change, from the wild to the tame.

Some dozen quartz-mills are now in full operation, testing and receiving the wealth from our numerous quartz lodes. The Gambrinus mill, after eighty hours' crushing rock, cleaned up $29,000 in amalgam. The first week's run of H. H. Raymond's mill on Granite Creek yielded fifty pounds. At Owyhee, the Morning Star mill, from the first ten tons of rock, extracted one ton of amalgam. Some of the richest ledges are the Pomona, Kentucky, Allen, Goldhill, Benton, Chickahominy, and Astor.

Young Idaho, when thoroughly developed, will be the Queen of Wealth on our little slope of the Rocky Mountains. Her little fertile valleys and grassy plains will furnish homes for thousands of people for ages to come. Her numerous water-privileges are excellent and available. By-the-way, the Vallisco Water Co. are digging two ditches, which are estimated to carry five thousand inches of water each. One of them begins near the upper end of Beaver Dick's ranch, on the northeast side of Boise River. This ditch is designed to furnish Boise valley with water for mechanical, agricultural, mining, and city purposes. The other ditch is the same size as the first named, and the water is to be taken out at Rocky Point. It will run down the valley between the fort and Boise City, and will furnish water to irrigate the whole valley on the north side of the river, and water for mills, &c. We are glad the work is rapidly approaching completion.

Below is a note of distances from Boise City to different mining camps :—

From Boise City to Owyhee, sixty-five miles ; to Placerville, Centreville, Pomona, Idaho City, thirty-five miles, N. E. ; to Silver City, Crooked River, Yuba Creek, Middle Boise, seventy miles, N. E. ; to South Boise, Rocky Bar, Happy Camp, eighty miles, E. ; to Salt Lake City, two hundred and ninety miles, E.

Centreville is making preparations for winter amusements: sleighs are being built, and a dancing-school is being organized. It is a pretty fast place, yet very quiet.

The Pioneer Ditch Company are extending their ditch down Grimes's Creek, which will cover plenty of good mining-ground. The miners, as a general thing, are "holed up" for the winter, excepting when a good quartz lode is discovered, when out they rush, armed to the teeth with picks and shovels, and charge around feet.

Pomona is a new town, just located by John Wallace & Co., and lies between Centreville and Idaho City, just below the junction of Gold Channel with Quartz Gulch. The elevated mounds around Pomona are dotted with cabins of miners who are engaged in searching for hidden wealth.

Mining on Elk Creek and Moore's Creek is about suspended on account of the creeks breaking in and filling up the diggings. William F. Glenn & Co. estimate their damages at $500, and other claims below are badly damaged.   J. F. J.

OWYHEE BULLION.

T. J. Carter, Esq., of this city, has received a quantity of crude bullion from Silver City, Owyhee, brought by Mr. Hotop, and deposited for assay at Tracy & King's. This was from ore crushed at A. P. Minear & Co.'s mill, and is another evidence of what Owyhee will produce in the way of Silver Bricks, the coming season. There are nearly five hundred ounces in the lot, and we are pleased to see it arriving in such liberal shipments, as there is no telling how many fine brick buildings will rise up in consequence of the productiveness of this polished material. Matters are progressing very handsomely in Owyhee. At the annual meeting of the Morning Star Gold and Silver Tunnel Company, Hon. A. C. Swift was elected President and A. P. Minear, Esq., Secretary. The Trustees of the Company are: Governor Gibbs, of our city; Hon. A. C. Swift, of Silver City; W. S. Bennington, A. P. Minear, and Uriah Perry. The tunnel of this Company is being pushed forward night and day, into the very comb-work of one of the richest mountains of ledges ever known, and promises the most abundant wealth to its projectors and every person interested in it. We inadvertently mis-stated the dimensions of this important work now many weeks since, and take this occasion to correct the same— the tunnel is four and a half by six and a half feet in the clear There are hundreds of ledges within a very short distance of the present extension, the owners of which have abandoned the sinking of shafts, preferring to use the tunnel. The number of new ledges that will be struck in this enterprise is, of

6*

course, unknown to man, but they must be innumerable, from the character of the region through which it passes. Owyhee is a busy place just now, and one year more will show an immense increase in its bullion shipments.

IDAHO.

MESSAGE OF GOVERNOR LYON.

Governor Lyon's message to the Territorial Legislature of Idaho is a patriotic and congratulatory document. We extract that portion of it referring to the mining and agricultural interests of the Territory:—

* * * I congratulate you upon the new and important discoveries of the precious metals, from the mountains of Pend d'Oreille, the gulches of South Boise to the shores of the Bruno, rewarding well the arduous zeal of the pioneer and the prospector. * * * I would respectfully submit to you for your consideration the mining interests of this Territory. By judicious legislation you will invite outside capital, without stint, to aid in their development; this interest should be most kindly fostered as the bed-rock of Idaho's prosperity. Some system of general legislation, by which security of the claimants of claims, and uniform laws throughout the different mining camps could be had, would accomplish an end earnestly desired. In this connection I would call your attention to the gold and silver quartz lead law, and for the better protection of that hardy class of our population to whom danger is not a sentiment and fear is unknown, who, amid trackless wastes, snow-clad mountains and deep-down cañons, pursue their enterprises with no capital but their rough hands, and no defender but their revolvers; who, without the knowledge of books, have given the world its choicest geological, geographical, and agricultural information, in tracing ledges, traversing rivers, and exploring valleys, which, but for them, would still have remained an unknown land beyond the Rocky Mountains. That

under no circumstances should they lose all the benefits resulting from their original discoveries, legislation should take the
place of litigation, and proper amendments be made which
would secure to them all their hard-earned rights and privileges. The vast unmeasurable structural wealth embodied
within our confines, so nicely balanced of mineral, farming, and
grazing interests, with mountain forests of timber-land, and
water-power of every description, eminently adapt us for a self-
supporting community. The fertile bottom-lands of the St.
Joseph, Cœur D'Alene, Spokane, La-toh, Palouse, Lapwaii,
Koos-koos-kia, Nas-so, Payette, Weiser, Boise, Malade, and
their tributaries, would alone sustain, properly cultivated, a
population larger than most of the Atlantic States; while the
ranges of nutritious "bunch-grass," suitable for herds, cover
millions of acres. Add to this placer diggings, of greater or
less richness, extending for hundreds of square miles, with well-
defined gold and silver bearing quartz ledges, unrivalled by
those of Mexico or Peru; a glorious climate, with Syrian sum-
mers and Italian winters, bespeak the permanence of our un-
told resources, and the prosperity that surely and positively
awaits their development.

THE OWYHEE MINES.

JORDAN CREEK, August 5, 1864.

Since the world has some interest, at least perspective if not
immediate, in what is generally known as the "Owyhee" silver
mining region, and as the world looks to the "Union" for its
full share of general information, and as that paper, although a
favorite of our reading population, seems to have no regular correspondent here, I propose to furnish a few items, to give at
least some general idea of the present condition of our section
of Idaho Territory. Situated in the southeast corner of the
Territory of Idaho, we feel more affinity than the remainder of
the Territory for both Nevada and California. Our news is

furnished from that direction, and should Red Bluff, as is asserted, prove not much more distant from us than the Columbia, and the road which has been already incorporated and commenced be made practicable, then California may look for a most important addition to her trade in the direction of Jordan Creek, which, by-the-by, flows some fifty miles from our mines, northwesterly, before emptying into the Owyhee, though the latter seems to be the popular name for this mining region.

It was one year ago that the miners on Jordan Creek made discoveries of gold and silver bearing quartz, and the Morning Star and Oro Fino ledges were located, the first scarce known to contain any thing of value, and the last showing croppings excessively rich in gold of a low fineness. The first assay made showed over $7,000, mostly gold; while the Morning Star, when tested, went $3,400, mostly silver. And these figures set the capitalists and speculators of Idaho, and also of Oregon, on the *qui vive*, with a severe attack of " quartz on the brain." Portland, ambitious of achieving every variety of commercial and financial distinction, manifested her faith in Owyhee by the establishment of her stock exchange; and the remainder of Idaho, seemingly jealous of the start we are gaining in the world, either based upon our independence of her other commercial centres, or of other quartz interests suffering in comparison, turns the cold shoulder upon us and sneeringly discredits our claims. We have no doubt that the Boise Basin and South Boise have strong claims on the world at large, and that their quartz will prove rich and profitable enough ; and as we shall, in another month, have three quartz-mills in operation, and as some of our best ledges are being thoroughly prospected in the mean time, we are content "to labor and to wait" until these mills can speak for us.

Strange to say, while ledges have been discovered almost by hundreds, and many with the richest of croppings both in gold and silver, none have been found that to-day stand so well in the public mind as the two I have named; and stranger, too, the "Oro Fino" is about the eastern boundary, and the "Morning Star" the western limit of the mining district, with a perfect network of quartz threading and interlacing through two miles that lie between, the whole range being scarce more than that distance wide, and some three miles in length. I am told by J. A. Chittenden, who has lately established an assay-office here, that while this region much resembles Reese River in character, he has never seen such rich rocks elsewhere as are shown him here. The Oro Fino, Morning Star, Whiskey, Home Ticket, Alabama, Mahala, Potosi, Golden Chariot, Silver Legion, Allison, Nondescript, and others, have very rich silver rock—sometimes almost marbled with gold; while there are several smaller ledges that are excessively rich in gold of low fineness.

At the present time several ledges are being prospected with the most favorable results. Tunnels are being run on the Oro Fino and the Allison. The first has run in on the ledge over a hundred feet, and finds from four to six feet of quartz, and takes out rock of great value. Some ground has been sold at $100 per foot, but cannot be readily bought. The Allison Company are reported to have struck their ledge, found good pay, and from eight to ten feet of quartz, but the thing is kept mum by those interested, who are looking for a few more feet.

On the Morning Star in several places, the Crane and Driggs, Oro Fino, the Silver Legion, Lady Candis, Home Ticket, War-Eagle, Nondescript, Ham-Fat, Noonday, Potosi, and Niagara, shafts are being sunk, and more or less work is being done on very many others, in many cases to me unknown, and with very general satisfaction, too; and a few companies who have money are going on with their work in a manner that shows their confidence in its results.

Very few ledges have been opened to any considerable extent, and I have knowledge of but few of these, having my own interests to secure; and I write at present not to hoist some man's ledge into notoriety or to advance private interests, for I have endeavored to leave out the ground I am interested in the most, and seek to afford the world general information of our prospects and resources, and refer them to the results of the running of our mills to bear me out and establish beyond a question the wealth and importance of this mining district. I wish more particularly to assert that our ledges invariably improve in width and richness, and my belief that there are ledges crowded out of memory now by richer veins, mere spurs of themselves, that will be found very wide and will prove to be of the most permanent value. The country is scarce prospected, and other districts have been discovered, ranging from ten to twenty-five miles distant, that in time will prove formidable rivals to this, and where the ledges furnish rock that assay well, and are large enough to satisfy anybody.

GENERAL MATTERS.

The Oro Fino Mountain, on which many of the best ledges are situated, is very high, being some two thousand feet above the level of Jordan Creek, and overlooks the dreary region through which Snake River winds like a thread of silver, but much needing an emerald setting of green banks and shading boughs. Of timber we have enough for many years, as our ledges will not exhaust the lumber-yards, as does the famous Comstock, and the mountain-sides are abounding in mountain mahogany and juniper, excellent for firewood. Water-power is unreliable, but the mountain-springs are cool and pure, and perhaps conduce much to the excellent health that prevails. Money is scarce, times are hard, and business consequently dull. No doubt this is an excellent time for investment, better than may occur again during a century, for I have faith in the future of Owyhee, and that the croppings of our mountain-sides are but a small token of the wealth that lies below.

Across Snake River to the east, distant some sixty miles in the Boise valley—far the choicest spot among the mountains, and within fifteen miles of the mines down Jordan Creek—is a beautiful valley that will raise ample supplies for home consumption.

But we have one great drawback, and must appeal to the General Government for protection from the savages that surround us, and from whom we have suffered much, and who have lately been chastised only at great private expense, and the loss of some of the most valuable lives among us. Of course this will prove a serious evil, and prevent the early settlement and thorough prospecting of a section that promises to contribute its full share toward the national wealth and prosperity.

MINING LANDS OF IDAHO TERRITORY.

AN ACT CONCERNING CORPORATIONS.

Be it enacted by the Legislative Assembly of the Territory of Idaho as follows: SECTION 1. Corporations for manufacturing, mining, mechanical, chemical, or agricultural purposes, for constructing telegraph-lines, for making roads, for establishing ferries, for building bridges, for conveying water, or for the purpose of engaging in any species of trade or commerce, may be formed according to the provisions of this Act. Such corporations and members thereof being subject to the conditions and liabilities herein imposed, and to none other: *Provided,* That nothing in this section shall be so construed as to authorize a company formed under it to own or hold possession of more than fourteen hundred and forty acres of land, or to authorize an individual member of such company or association, in his corporate capacity, to hold, own, or possess a number of acres to exceed eighty; and, *Provided further,* That no corporation formed for agricultural purposes shall be allowed to hold any mineral lands under the provisions of this act.

SEC. 2. Any three or more persons, who may desire to form a company for any one or more of the purposes specified in the preceding section, may make, sign, and acknowledge, before some officer competent to take the acknowledgment of deeds, and file in the office of the County Clerk, or Clerk of the District Court of the Judicial District in which the principal place of business of the company is intended to be located, and a certified copy thereof under the hand of the clerk, and seal of said court, in the said district, in the office of the Secretary of the Territory, a certificate in writing, in which shall be stated the corporate name of the company, the object for which the company shall be formed, the amount of its capital stock, the time of its existence, not to exceed fifty years, the number of shares of which the stock shall consist, the number of trustees and their names, who shall manage the concerns of the company for the first three months, and the name of the city or town and county in which the principal place of business of the company is to be located.

SEC. 3. A copy of any certificate of incorporation filed in pursuance of this Act, and certified by the County Clerk, or the Clerk of the District Court in the county or district in which it is filed, or his deputy, or by the Secretary of the Territory, shall be received in all courts and places as presumptive evidence of the facts therein stated.

SEC. 4. When the certificate shall have been filed, the persons who shall have signed and acknowledged the same, and their successors, shall be a body politic and corporate, in fact and in name, by the name stated in the certificate; and by their corporate name shall have succession for the period limited, and power—1. To sue and be sued in any court. 2. To make and use a common seal, and alter the same at pleasure. 3. To purchase, hold, sell, and convey, such real and personal estate as the purposes of the corporation shall require. 4. To appoint such officers, agents, and servants, as the business of the corporation may require; to define their powers, prescribe their duties, and fix their compensation. 5. To require of them such security as may be thought proper for the fulfilment of their duties, and to remove them at will, except that no trustee shall be removed from office unless by a vote of two-thirds of the whole number of trustees, or by a vote of a majority of the trustees, upon a written request signed by stockholders legally representing two-thirds of the whole stock. 6. To make by-laws not inconsistent with the laws of this Territory, or of the United States, for the organization of the company, the management of the property, the regulation of its affairs, the transfer of its stock, and for carrying on all kinds of business within the objects and purposes of the company.

SEC. 5. The corporate powers of the corporation shall be exercised by a board of not less than three trustees, who shall be stockholders in the company, and a majority of them citizens of the United States, and residents of this Territory, and who shall, after the expiration of the term of the trustees first elected, be annually elected by the stockholders, at such time and place, and upon such notice, and in such mode as shall be directed by the by-laws of the company; but all elections shall be by ballot, and each stockholder, either in person or by proxy, shall be entitled to as many votes as he owns shares of stock, and the persons receiving the greatest number of votes shall be and act as trustees. When any vacancy shall happen among the trustees, by death, resignation, or otherwise, it shall be filled for the remainder of the year in such manner as may be provided by the by-laws of the company.

SEC. 6. If it should happen, at any time, that an election of trustees shall not be made on the day designated by the by-laws of the company, the corporation shall not, for that reason, be dissolved, but it shall be lawful, on any other day, to hold an election for trustees in such manner as shall be provided for by the by-laws of the company; and all acts of trustees shall be valid and binding upon the company until their successors shall be elected.

SEC. 7. A majority of the whole number of trustees shall form a board for the transaction of business, and every decision of a majority of the persons duly assembled as a board shall be valid as a corporate act.

SEC. 8. The first meeting of the trustees shall be called by a notice signed by one or more of the persons named trustees in the certificate, setting forth the time and place of the meeting, which notice shall be either delivered personally to each trustee or published at least ten days in some newspaper of the county in which is the principal place of business of the corporation; or, if no newspaper be published in the county, then in some newspaper published nearest thereto.

SEC. 9. The stock of the company shall be deemed personal estate, and shall be transferable in such manner as shall be prescribed by the by-laws of the company; but no transfer shall be valid except between the parties thereto, until the same shall have [been] so entered on the books of the company as to show the names of the parties by and to whom transferred, the number and designation of the same, and the date of the transfer.

SEC. 10. The trustees shall have power to call in, and demand from the stockholders, the sums by them subscribed, at such times and in such payments or instalments as they may deem proper. Notice of each assessment shall be given to the stockholders personally, or shall be published once a week for at least four weeks in some newspaper published at the place designated as the principal place of business of the corporation, or if none is published there, in some newspaper published nearest to such place. If, after such notice shall have been given, any stockholder shall make default in the payment of the assessment upon the shares held by him, so many of such shares may be sold as will be necessary for the payment of the assessment on all the shares held by him. The sale of said shares shall be made as prescribed in the by-laws of the company; *Provided*, That no sale shall be made except at public auction to the highest bidder, after a notice of thirty days, published as above directed in this section; and that at such sale the person who will agree to pay the assessment so due, together with the expense of advertisement and the other expenses of sale, for the smallest number of whole shares, shall be deemed the highest bidder.

SEC. 11. Whenever any stock is held by any person as executor, administrator, guardian, or trustee, he shall represent such stock at all meetings of the company, and vote accordingly as a stockholder.

SEC. 12. Any stockholder may pledge his stock by a delivery of his certificate or other evidence of his interest, but may, nevertheless, represent the same at all meetings, and vote accordingly as a stockholder.

SEC. 13. It shall not be lawful for the trustees to make any dividend except from the surplus profits arising from the business of the corporation, nor to divide, withdraw, or in any way pay to the stockholders, or any of them, any part of the capital stock of the company ; nor to reduce the capital stock, unless in the manner prescribed in this Act; and in case of any violation of the provisions of this section, the trustees under whose administration the same may have happened, except those who may have caused their dissent therefrom to be entered at large on the minutes of the board of trustees at the time, or were not present when the same did happen, shall, in their individual and private capacity, be jointly and severally liable to the corporation, and the creditors thereof, in the event of its dissolution, to the full amount so divided, withdrawn, paid out, or reduced ; *Provided*, That this section shall not be so construed as to prevent a division and distribution of the capital stock of the company which shall remain after the payment of all its debts, upon the dissolution of the corporation or the extinction of its charter.

SEC. 14. The total amount of the debts of the corporation shall not at any time exceed the amount of the capital stock actually paid in ; and in case of any excess, the trustees under whose administration the same may have happened, except those who have caused their dissent therefrom to be entered at large on the minutes of the board of trustees at the time, and except those not present when the same did happen, shall, in their individual and private capacities, be liable, jointly and severally, to the said corporation, and in the event of its dissolution, to any of the creditors thereof, for the full amount of such excess.

SEC. 15. No corporation organized under this act shall, by any implication or construction, be deemed to possess the power of issuing bills, notes, or other evidences of debt, for circulation as money.

SEC. 16. Each stockholder shall be individually and personally liable for his proportion of all the debts or liabilities of the company contracted or incurred during the time that he was a stockholder, for the recovery of which, joint or several actions may be instituted; and when a judgment in such action shall be recovered against joint stockholders, the court, on the trial thereof, shall apportion the amount of the liability of each, and in the execution thereof, no stockholder shall be liable beyond his proportion so ascertained.

SEC. 17. No person holding stock as executor, administrator, guardian, or trustee, or holding it as collateral security, or in pledge, shall be personally subject to any liability as a stockholder of the company; but the person pledging the stock shall be considered as holding the same, and shall be liable as a stockholder accordingly; and the estate and funds in the hands of the executor, administrator, guardian, or trustee, shall be liable in like manner, and to the same extent as the testator or intestate, or the ward, or person interested in the trust-fund would have been if he had been living, and competent to act and hold the stock in his own name.

SEC. 18. It shall be the duty of the trustee of every company incorporated under this act for the purpose of ditching, mining, or conveying water for mining purposes, to cause a book to be kept, containing the names of all persons, alphabetically arranged, who are or shall become stockholders of the corporation, and showing the number and designation of shares of stock held by them respectively, and the time when they respectively became the owners of such shares; also, a book or books, in which shall be entered at length, in a plain and simple manner, all by-laws, orders, and resolutions, of the company and board of trustees, and the manner and time of their adoption; which books, during the business hours of the day (Sundays, Fourth of July, and the Twenty-fifth day of December, excepted), shall be open for the inspection of stockholders and the creditors of the company, each individual stockholder, and their duly authorized agents and attorneys, at the office or principal place of business of the company; *Provided,* That the office and books of every such company shall be kept, and the books of the company shall be open as aforesaid, in the county in which their principal business is transacted; and every stockholder and creditor as aforesaid, or their agents or attorneys, shall have the right to make extracts from such books, or upon payment of reasonable clerk's fees therefor, to demand and receive from the clerk, or other officer having the charge of such books, a certified copy of any entry, [which] shall be presumptive evidence of the facts therein stated in any action or proceeding against the company, or any one or more of the stockholders.

Sec. 19. If the clerk or other officer having charge of such books shall make any false entry, or neglect to make any proper entry therein, or shall refuse or neglect to exhibit the same, or allow the same to be inspected, or extracts to be taken therefrom, or to give a certified copy of an entry therein, as provided in the preceding section, he shall be deemed guilty of a misdemeanor, and shall forfeit and pay to the party injured a penalty of one hundred dollars, and all damages resulting therefrom to be recovered in any court of competent jurisdiction in this Territory; and for neglect to keep up such books for inspection, and in the place provided for in the last section, the corporation shall forfeit to the people of Idaho Territory the sum of two hundred and fifty dollars for every day they shall so neglect, to be sued for and recovered before any court of competent jurisdiction in the county or district in which the principal business of such company is transacted; and it shall be the duty of the district attorney within and for such district, to prosecute such action in the name of, and for the benefit of, the people of Idaho Territory; and it is further *Provided*, That in case any such incorporated company shall refuse or neglect, for the space of one full year after the passage of this Act, to comply with the provisions of this and the preceding section, then, upon the showing of such facts, by petition of any person aggrieved thereby, and due proof thereof before the district judge in the district in which said company's principal business is transacted, after such company shall have been duly notified thereof by summons, to be issued by said judge, citing such company to appear before said judge at a time and place therein mentioned, which shall not be less than ten, nor more than thirty days from the date of such summons, such company shall by such judge be declared and decreed to be disincorporated so far as to deprive said company of all the privileges of this Act, but in no manner to affect the remedy of all persons against such company, to be exercised as this Act provides; *Provided*, That nothing contained in the provisions of this section concerning the disincorporating of such companies, shall be so construed as to prevent the enforcement of the other remedies in this section mentioned, at any time after the passage of this Act, except as herein provided.

SEC. 20. Any company incorporated under this Act, may, by complying with the provisions herein contained, increase or diminish its capital stock to any amount which may be deemed sufficient and proper for the purposes of the corporation; but before any corporation shall be entitled to diminish the amount of its capital stock, if the amount of its debts and liabilities shall exceed the sum to which the capital is proposed to be diminished, such amount shall be satisfied and reduced so as not to exceed the diminished amount of capital.

SEC. 21. Whenever it is desired to increase or diminish the

amount of capital stock, a meeting of the stockholders may be called by a notice signed by at least a majority of the trustees, and published for at least four weeks in some newspaper published in the county where the principal place of business of the company is located, or in some newspaper nearest thereto; which notice shall specify the object of the meeting, the time and place where it is to be held, the amount to which it is proposed to increase or diminish the capital; and a vote of two-thirds of all the shares of stock shall be necessary to an increase or diminution of the amount of capital stock.

SEC. 22. If at any meeting so called a sufficient number of votes has been given in favor of increasing or diminishing the amount of capital, a certificate of the proceedings, showing a compliance with these provisions, the amount of capital actually paid in, the whole amount of the debts and liabilities of the company, and the amount to which the capital stock is to be increased or diminished, shall be made out, signed and verified by the affidavit of the chairman and secretary of the meeting, certified by a majority of the trustees, and filed as required by the second section of this Act; and when so filed, the capital stock of the corporation shall be increased or diminished to the amount specified in the certificate.

SEC. 23. Upon the dissolution of any corporation formed under this Act, the trustees, at the time of the dissolution, shall be trustees of the creditors and stockholders of the corporation dissolved, and shall have full power and authority to sue for, and recover the debts and property of the corporation by the name of trustees of such corporation, collect and pay the outstanding debts, settle all its affairs, and divide among the stockholders the money and other property that shall remain after the payment of the debts and necessary expenses.

SEC. 24. Any corporation formed under this Act may dissolve and disincorporate itself by presenting to the district judge of the district, in which the meeting of the trustees is usually held, a petition to that effect, accompanied by a certificate of its proper officers, and setting forth that at a general or special meeting of the stockholders, called for that purpose, it was decided by a vote of two-thirds of all the stockholders to dis-incorporate and dissolve the corporation. Notice of the application shall then be given by the clerk, which notice shall set forth the nature of the application, and shall specify the time and place at which it is to be heard, and shall be published in some newspaper of the county once a week for four consecutive weeks, or if no newspaper is published in the county, by advertisement posted up for thirty days in three of the most public places in the county. At the time and place appointed, or at any other to which it may be postponed by the judge, he shall proceed to consider the application, and if satisfied that the corporation has taken the necessary preliminary steps, and obtained the necessary vote to dissolve itself, and that all claims against the corporation are discharged, he shall enter an order declaring it dissolved.

Sec. 25. This act shall take effect and be in force from and after the date of its approval by the Governor.

Passed the House of Representatives, December 28th, 1864.

JAMES TUFTS,
Speaker of the House of Representatives.

Passed the Council, December 24th, 1863.

JOSEPH MILLER,
President of the Council.

Attest:

JAMES McLAUGHLIN,
Chief Clerk Council.

Approved, the 4th day of January, 1864.

WILLIAM B. DANIELS,
Acting-Governor of Idaho Territory.

I hereby certify that the foregoing law contained in this printed paper is a true and literal copy of the enrolled law, passed by the First Legislative Assembly of said Territory, held during the months of December, eighteen hundred and sixty-three, and January and February, eighteen hundred and sixty-four, on file in my office.

Witness my hand and the seal of the Territory here-
[SEAL.]   unto annexed, this seventeenth day of September, eighteen hundred and sixty-four.

SILAS D. COCHRAN,
Acting-Secretary of the Territory.

———

We could ask no better evidence of the great richness of the mines of Idaho, and their prospective brilliant future, than the following extracts from an editorial of the San Francisco Daily *"Alta California"* of January 7, 1865 :—

"IDAHO AND THE ROAD THITHER.

"The mines of southwestern Idaho are rapidly increasing in importance.

"The districts of Idaho City, South Boise, and Owyhee, have taken a high position among the treasure supplies of the world. The estimated production of the Territory last year was six millions of dollars, thus placing it among the principal gold-fields. The growth of the country has been very rapid, and so also the increase in the yield of the precious metals. The following figures show the deposits of Idaho gold, made at the San Francisco mint in 1864 :—

	OUNCES.
January	10,366 55
February	6,473 54
March	7,247 98
April	5,649 44
May	6,444 85
June	28,442 95
July	27,120 12
August	38,892 28

7

	OUNCES.
September	22,134 72
October,	25,748 61
November	38,363 68
December	245 31

"The sum deposited in December was very small, because the mint closed on the third of that month and did not open again until after New Year. It will be observed that there was a rapid increase from the beginning to the end of the year. The average of the last four months before December is four times as great as that of the first four months.

"The metal varies in fineness from 500 to 800 fine in gold, the average being about 760 in gold, 230 of silver, and 10 of base metals. The alloy is nearly all silver, and in this respect, and in the large quantity of silver, the metal bears a remarkable resemblance to that found at Gold Hill before the Comstock lode was opened. At Idaho, as at Gold Hill, the gold is gradually decreasing in fineness as the mines go down—the placer gold found in deep diggings containing more silver than that found near the surface. These are indications of rich and permanent silver-mines, of which a number have already been opened. The value of gold 1,000 fine, is about $20 per ounce; of 500, is $10 per ounce for the gold and 75c. for the silver; and of 760 fine, about $15 50 per ounce for gold and silver together. At that rate the Idaho gold deposited in the San Francisco mint last year was worth about $3,400,000. Large quantities of gold from Idaho, as well as of that from California, were not sent to the mint.

"The population of Idaho is about 20,000, and is rapidly increasing. The country is rich, and the people wish to import many comforts and luxuries, as well as to export their silver and gold.

"What road shall they take? In every case they will come to San Francisco, but shall they come by way of the Sacramento or the Columbia River? Can any thing be done to shorten the route and decrease the expense of freight and passage? These are questions which deserve the attention of the capitalists and merchants of California.

"At first thought the presumptions are all in favor of the Columbia River route. The mines are in the basin of that river, and ever since they were opened, all the trade has gone that way. Nevertheless, when we examine more closely, we discover many advantages in favor of the Sacramento route.

*   *   *   *   *   *   *   *   *   *   *

"From Red Bluffs there is now a good road to Susanville, in Honey Lake Valley. From that point to Pueblo, which is a mining district, 80 miles north of Humboldt City and 40 miles north of Black Rock, there is a good road, with a few bad places where a good road might be made at a slight expense. From Pueblo to Owyhee the distance is 100 miles; the intervening country being mountainous and bare of timber, but rich in grass, and well supplied with water. To construct a good wagon-road over this distance would cost, it is estimated, $30,000, and then the road to California would be open at once. There are already good roads from Owyhee to Boise City, Idaho City, and South Boise."

TERRITORY OF MONTANA.

The Territory of Montana, contiguous as it is to Idaho, is a region replete with interest to the miner, geologist, and capitalist. Professor Eaton, of New York, has examined quite thoroughly this section of our mining region, and any evidence from him is looked for with marked interest.

Two years ago this winter the first party of miners engaged in the successful working of the gold-mines within the present boundaries of Montana Territory wintered at Bannock City, on a creek called by them Grasshopper. This creek is designated upon the maps appended to the "History of the Expedition under the Command of Captains Lewis and Clark to the Sources of the Missouri, performed during the Years 1804–5–6," as Willard's Creek, after one of their men, Alexander Willard. Its location is on about the forty-fifth parallel of latitude, and one hundred and twelfth degree of longitude west from London. Gold was first discovered on this creek in the month of August, 1862. It was found in the bed of the creek, and beneath the alluvial soil that formed its banks, in the sand and gravel which covered the bed-rock, at a depth varying from three to twenty-five feet. As the population increased, and they extended their researches further, they found the shining metal upon the high bluffs which walled the stream on either side; and prospecting still further, they found it interspersed in the veins of quartz and rock running through the high and low hills through which the stream flows. The country at the time of the discovery was in the wildest state of nature. The nearest point where supplies of provisions, working-utensils, and clothing, could be obtained, was Great Salt Lake City, distant four hundred and thirty miles. The nearest habitation of a white man was at Box Elder, three hundred and fifty miles.

The miners were compelled to improvise such rude appliances for digging and washing the gold as their retired position from the haunts of civilization would permit. It was found in such bountiful quantities, however, that, notwithstanding the rude means they had to work with, they could collect from ten dollars to one hundred dollars per day, and frequently as high as six hundred dollars per day were washed out with a single rocker. The quartz lodes in the surrounding region have proved to be of sur-

passing extent and richness.   One of them is probably one
of the richest gold-mines ever discovered in the northwestern territory.   At a distance of sixty-five feet in this
mine the ore has a red, burnt appearance, Nature having
apparently performed the work of desulphurization her-
self.   The discoverers of this lode, in the winter of 1862
and 1863, pounded the ore in a stone mortar, and made
from ten dollars to twenty dollars per day in the operation.

Recent discoveries have been made in silver which
promise to make Montana one of the richest silver-pro-
ducing Territories in the United States.   These silver lodes
are in the immediate vicinity of Bannock City, and upon
the Rattlesnake Creek, distant some thirteen miles.   Discoveries have also been made upon the Ram's-Horn
Gulch, in both gold and silver, and in various localities
on the waters of the Jefferson fork of the Missouri.   The
lodes at Bannock and on the Rattlesnake have been more
fully prospected than those of any other locality, and
their immense yield, by actual assay, is almost fabulous.
A well-known scientific gentleman of New York, Professor A. K. Eaton, visited Montana Territory in the
interest of some parties in this city, and spent some six
months in the country during the last summer and fall.
Before his arrival the existence of silver was unknown in
the Territory, although its presence was suspected.

Various assays have been made from these ores, which
exhibit a yield of from eighty dollars to four thousand
five hundred dollars per ton of two thousand pounds,
some of them showing forty per cent. of gold on the net
amount of the assay.

The country in the vicinity of these mines is well supplied with wood and water, and in this respect possesses

all the advantages necessary for the cheap reduction of the ores.

———

THE NEW MINING LAW OF MONTANA TERRITORY.

Be it enacted by the Legislative Assembly of the Territory of Montana :—

SECTION 1. That any person or persons who may hereafter discover any quartz lead, lode, or ledge, shall be entitled to one claim by right of discovery, and one claim each by pre-emption.

SEC. 2. That in order to entitle any person or persons to record in the County Recorder's office of the proper County, any lead, lode, or ledge, either of gold or silver, or claim thereon, there shall first be discovered on said lode, lead, or ledge, a vein or crevice of quartz ore, with at least one well-defined wall.

SEC. 3. Claims on any lead, lode, or ledge, either of gold or silver, hereafter discovered, shall consist of not more than two hundred feet along the lode, lead, or ledge, together with all dips, spurs, and angles, emanating or diverging from said lead, lode, or ledge, as also fifty feet on each side of the centre of said lead, lode, or ledge, for working purposes : *Provided*, That when two or more leads, lodes, or ledges, shall be discovered within one hundred feet of each other, either running parallel or crossing each other, the ground between such leads, lodes, or ledges, shall belong equally to the claimants of the said leads, lodes, or ledges, without regard to priority of discovery or pre-emption.

SEC. 4. When any leads, lodes, or ledges, shall cross each other, the quartz ore or mineral, in the crevice or vein at the place of crossing, shall belong to, and be the property of the claimant upon the lead, lode, or ledge first discovered.

SEC. 5. That before any record shall be made under the provisions of this Act, there shall be placed at each extremity of the discovery claim, a good and substantial stake, not less than five inches in diameter, said stake to be firmly planted or sunk into the ground, extending two feet above ground ; that upon each stake there shall be placed in legible characters, the name of the lead, lode, or ledge, and that of the discoverer or discoverers, the date of the discovery, and the name of each pre-emptor or claimant, and the directions or bearings, as near as may be, of his or her claims. Said stakes, and the inscriptions thereon, to be replaced at least once in twelve months by the claimants on said lode, lead, or ledge, if torn down or otherwise destroyed.

SEC. 6 Notice of the discovery or pre-emption upon any lead, lode, or ledge, shall be filed for record in the County Recorder's office of the county in which the same may be situated, within fifteen days of the date of the discovery or pre-emption, and there shall, at the same time, be an oath taken before the recorder, that the claimant or claimants are, each and all of them, *bona fide* residents of the Territory of Montana; and there shall be deposited in the recorder's office, either by the discoverer or some pre-emptor, a specimen of the quartz ore or mineral extracted or taken from said lead, lode, or ledge, which said specimen shall be properly labelled by the recorder, and preserved in his office.

SEC. 7. That any person or persons who shall take up or destroy, or cause the same to be done, any of the said stakes, or who shall, in any wise, purposely deface or obliterate any part or portion of the writing or inscriptions placed thereon, shall be deemed guilty of a misdemeanor, and upon conviction thereof before any court of competent jurisdiction, shall be punished by a fine of not more than one thousand dollars, or by imprisonment in the county jail not more than ninety days, or by both such fine and imprisonment.

SEC. 8. That the amount of ground which may be taken up on any lead, lode, or ledge, in addition to the discovery claim, shall be limited to one thousand feet along said lead, lode, or ledge, in each direction from the discovery claim thereon.

SEC. 9. All lead, lode, or ledge claims taken up and recorded in pursuance with the provisions of this act, shall entitle the person recording, to hold the same to the use of himself, his heirs and assigns; and conveyances of quartz claims shall hereafter require the same formalities, and shall be subject to the same rules of construction, as the transfer and conveyance of real estate.

SEC. 10. That if, at any time previous to the passage of this Act, claims have been taken up and recorded in the recorder's office of the proper county, upon any actual or proper lead, lode, or ledge of quartz ore or mineral, the owners or proper claimant or claimants of said respective claims shall hold the same to the use of themselves, their heirs and assigns.

SEC. 11. That the Act relating to the discovery of gold and silver quartz lodes, and the manner of their location, passed by the Idaho Legislature, and approved February 4th, 1864, and all other Acts or parts of Acts inconsistent with this Act, be, and the same are hereby repealed.

Sec. 12. This act to take effect and be in force from and after its passage.

I certify that the above is a true copy of an act passed by the Legislative Assembly of Montana, and approved December 26th, 1864.

SIDNEY EDGERTON, Governor.